BOATS OF THE AIR

BOATS OF THE AIR

An Illustrated History of Flying Boats, Seaplanes and Amphibians

David W. Wragg

ROBERT HALE · LONDON

First published in Great Britain 1984

ISBN 7090 1553 4

Robert Hale Limited
Clerkenwell House
Clerkenwell Green
London EC1R 0HT

Photoset and printed in Great Britain by
Redwood Burn Limited, Trowbridge, Wilts.
Bound by WBC Bookbinders Ltd.

Contents

Acknowledgements

I should like to thank those who assisted in the research and preparation of the material used in this book, and in particular those who assisted in the collection of the illustrations, many of whom are manufacturers who no longer have an interest in flying boat, amphibian or seaplane construction, and whose help reflects a real regard for aeronautical history. My thanks to: Fleet Air Arm Museum; Royal Air Force Museum; Imperial War Museum; National Maritime Museum; Musée de l'Air; Air New Zealand; Avions Marcel Dassault-Breguet Aviation; British Aerospace; Boeing; Canadair; de Havilland Canada; Dornier; Flight Refuelling; General Dynamics; Grumman Aerospace; Lufthansa; Martin Marietta; McDonnell Douglas; Qantas; Pan Am; Sikorsky Aircraft, and Short Brothers.

1
MORE THAN JUST AN IDEA

Some two-thirds of the earth's surface is covered by water, while not all of the land area is suitable for, or available for, the long runways of an airport. After all, there are real difficulties in siting airports on marsh or swamp, mountain or jungle, or simply because of urban development. Then again, many of the world's major cities are either on the water or close to it, even in the case of a number of those inland. Many land-locked countries appear to be blessed with lakes. A substantial number of the major capital cities are either on the coast or close to a wide river, including European capitals such as Helsinki, Stockholm, Oslo, Copenhagen, Amsterdam, Dublin, Lisbon and Athens; Rome is close to the open sea, London and Paris are on major rivers, Berlin is close to a major lake and Bonn is by a river, leaving only Madrid and Brussels far from the coast and a suitable river. The picture is much the same in the New World, with Washington astride the Potomac River and the old Brazilian capital of Rio de Janeiro on the coast, even though Brasilia is inland; Montevideo, Buenos Aires and Santiago de Chile are all coastal towns. Although not capital cities, major cities such as New York and Barcelona owe their commercial and industrial importance to their proximity to the sea. Indeed, the world's major ports are themselves of such a size that they are major centres of population and industry.

The advent of the helicopter has removed many of the advantages of water take-off and landing, or at least the necessity for it in many cases. However, of all the different categories of flying machine, the helicopter was amongst the last to appear, and it is only during the last thirty years or so that the helicopter has been able to offer a significant performance for commercial or military use, while it remains an expensive machine to buy and to operate, and is noisy and difficult to fly compared to conventional aircraft. As a result, there are very few successful commercial helicopter air routes in the world today. Yet, aviation has enjoyed the greatest growth, especially in the uncertain early years, in those parts of the world in which surface transport has traditionally been the most difficult due to a harsh terrain. It was not for nothing that one of the world's oldest airlines operated over the jungles of Colombia.

Of course, there is another aspect to this. The early aeroplanes suffered from indifferent undercarriages and there were few prepared aerodromes available in advance of the new air services and public demand, slow at first, for air travel. As aircraft grew larger and operated over longer distances, often well away from the major industrial centres of the developed world, adequate airfields were long in appearing, and so too were robust undercarriages. Indeed, for many years, aerodynamic progress was inhibited by the materials available, and this also had some effect on the timing of the helicopter's arrival.

Technical reliability was yet another problem. If an engine failed, even on a multi-engined aircraft, there was little enough power available for the aircraft to reach an airfield, and emergency landings were the order of the day on the main air routes, with an aircraft often making any number of landings on a flight from London to Paris. Emergency landings were difficult enough at the best of times, but what if the major part of a flight lay over the sea? Or for that matter jungle in which the only available landing place would be a lake or a stretch of river? It was obviously best for all concerned, aircraft and occupants, if the aircraft could alight on water without damage, perhaps even to be repaired so that the flight could continue.

So, everything seemed to point to the need for an aeroplane to be able to land and take off from water, especially if it was intended to operate on the longer distance routes, during the early days. As one pioneer was to say, the 'flying boat takes its airport on its bottom', and while this was an over-simplification of the difficulty in finding suitable landing sites, even for marine or hydro-aeroplanes, it was certainly easier than finding a suitable location, clearing it, transporting men and materials to the site, and then building runways. The military were just as aware of the advantages of the seaplane and flying boat as any airline.

Yet, no one can really be sure just who first gave serious consideration to the idea of water take-offs and landings. It may have owed something during the early days not so much to practical considerations as to the visionaries who foresaw that one day man would sail across the skies rather as he did across the sea. Typical of this school of thought was the Jesuit priest, Father Francesco de Lana de Terzi, who, in 1670, designed an aerial carriage or ship, and the Brazilian priest, Gusmao, who designed a flying machine during the early years of the eighteenth century—both of these foresaw water take-offs. However, neither of them could be regarded as true pioneers, having made designs without attempting to build.

Rather more prophetic was the great British pioneer, Sir George Cayley, who enjoys the distinction of identifying many of the problems and solutions to flight, and who built and flew the first man-carrying gliders during the middle years of the nineteenth century. It was Cayley who suggested that large airships might need to alight on water, and it is perhaps worth remembering that the first of Count Ferdinand von Zeppelin's airships, the LZ.1, ascended from a floating hangar at Friedrichshafen, on Lake Constance, for its maiden twenty-minute flight in July, 1900. Yet, Cayley's heavier-than-air designs were all fitted with wheels. Another pioneer, the talented but tragic Frenchman, Alphonse Penaud, designed a monoplane amphibian, in 1876, with the assistance of his mechanic, Paul Gauchet, although this interesting and advanced design was never built.

Rather more came from the work of the Australian pioneer, Lawrence Hargrave, whose lasting claim to fame was the invention of the box-kite in 1893, the form of which was adapted by other later and more successful pioneers, notably the Voisin brothers in France during the early twentieth century. Working in New South Wales, near Sydney, Hargrave had visited England to demonstrate his box-kites to the Royal Aeronautical Society, leaving them some examples for further experiment. After such success with the kites, it was only natural that he should then seek to apply propulsion to a larger man-carrying kite. His first attempt at powered man-carrying flight was with a landplane and was unsuccessful, but Hargrave tried again with a second aircraft which used a steam-engine to drive four flappers within the box-kite bays or cells. It was effectively a tandem-wing biplane, with four floats at the wingtips so that the aircraft could operate from water. However, this aircraft also failed during an attempt at flight in 1896, not surprisingly since neither the steam-engine nor the flappers offered any real way forward in aeronautical development. Moving to a new waterfront home at Woollahra Point, Hargrave built his third full-sized design in 1902, utilizing a tubular steel fuselage and replacing the biplane mainplane (although in fact as a tandem-wing aircraft, mainplane and fuselage were of equal span) and tailplane of his earlier designs with a multiplane design based on the work of the Englishman, Horatio Phillips, undoubtedly hoping to improve lift by using this complex 'venetian blind' design. Certainly, the Phillips' design had demonstrated lifting qualities during tethered trials in England. However, Hargrave's third design, probably the first aircraft to be built with a large central float and two wingtip floats, failed to become airborne despite the excellence of much of Hargrave's and Phillips' work because once again the steam-engine proved inadequate for the job. A further attempt in March, 1906, this time with a petrol-engined aircraft, was also unsuccessful and at this stage Hargrave finally gave up his aeronautical experiments.

It is usual for historians to attribute Hargrave's inability to make the most of his designs to geographical isolation, being well removed from the main centres of aeronautical activity in Europe and the United States, but that ignores both the availability of journals and reports on developments and the fact that he was well aware of developments, as his adaptation of Phillips' work showed. He had also, as already mentioned, visited London, coming into contact with other pioneers and in fact receiving an enthusiastic welcome which could well have resulted in regular correspondence with many like-minded individuals. However, it is likely that Hargrave suffered from not having a suitably qualified and interested companion with whom he could discuss his ideas, and this inability to generate fresh thought must have contributed considerably to his failures. Then too, while comfortably off, he was finding the growing cost of his experiments a drain on his resources, particularly as his domestic commitments became more demanding. In short, in a different continent and with a deeper purse, Hargrave might have achieved much more.

Success also evaded another illustrious pioneer, the American astronomer, Samuel Pierpoint Langley, who could hardly have claimed a shortage of funds since he was in receipt of what was, by the standards and values of the day, massive official aid.

Langley had enjoyed the benefit of many years of experiment. Starting in 1887, he conducted a number of whirling arm experiments – using the crude device which was commonly used to test aerofoils during the years before wind tunnels were invented – before building a total of some forty rubber-powered models, many of which were extensively modified as a result of testing. These models were followed during 1892–94 by the larger steam-powered models which Langley named, rather inaccurately and misleadingly, as Aerodromes, and which were numbered, again misleadingly, 0 to 6. The Aerodromes were unsuccessful at first, suffering from a number of problems including a lack of stability and wing distortion while in flight, in addition to difficulties with the launching mechanism, which Langley had installed on the roof of a houseboat moored on the Potomac River. Far from being disappointed with the results of this work, Langley rebuilt his Aerodrome No. 5 to a tandem-wing configuration, which he had discovered studying the work of, and by visiting, the Englishman, D. S. Brown. Aerodrome No. 4 enjoyed a similar reconstruction, becoming No. 6. These models were successful, with No. 5 flying for 3,300 feet on 6th May, 1896, and No. 6 for some 4,000 feet on 28th November, 1896.

Langley's work impressed the United States war department, which awarded him a $50,000 contract for the design and construction of a man-carrying aeroplane. Langley in his turn subcontracted the task of building a suitable engine to Stephen M. Balzer, who was asked to build a 12-hp petrol engine of not more than 100 lbs in weight, although eventually, after the engine was built and tested, Langley's associate, Charles Manly, had to redesign and rebuild the 5-cylinder rotary engine, which could only deliver 8 hp, to a radial engine. Meanwhile, Langley built a one-quarter scale model of his aircraft, which he had designated the *Aerodrome A*, with a 3.2-hp Manly–Balzer engine, although this model failed to fly on a straight and level course at first and modifications took until August, 1903.

In spite of these difficulties, Langley's confidence in his *Aerodrome A* remained undiminished. On completion in 1903, the full-sized aircraft incorporated Langley's tandem-wing configuration with equal span monoplane mainplane and tailplane, and with the pilot sitting within the fuselage, level with the front wing or mainplane and in front of the engine, which drove two pusher propellers. Control of the aircraft was primitive, with Langley being a supporter of the so-called 'chauffeur' school of thought, which failed to appreciate the three-dimensional nature of flight, leaving the pilot with just rudder and engine control, with the rest left to what designers such as Langley hoped would be adequate stability. Of course, such stability was fine with models, but it was a serious drawback for an aircraft which planned to change direction during flight. Even though his launching system had also discredited itself with his models, Langley also arranged for the full-sized aircraft, with Charles Manly as pilot, to be catapulted off the Potomac River houseboat, although the mechanism was modified to fit under rather than over the aircraft.

On 7th October, 1903, the *Aerodrome A* was launched from the houseboat but fouled the launching mechanism and crashed into the river. A second attempt, on 8th December, was simply a repeat performance, even to giving the unfortunate Charles Manly another unwanted ducking.

However, Langley had his admirers, in addition to the United States war department. A Russian of German origin and Austrian nationality, Wilhelm Kress, published a design for a floatplane in 1880, intending this to employ some form of propulsion. Several years later, in 1898, Kress actually built a full-sized floatplane, designed to carry a pilot but no passengers and powered by a 30-hp Daimler petrol engine driving two propellers set on the mainplane of the aircraft – a tandem-wing design similar to, and based upon, Langley's Aerodromes 5 and 6.

Samuel Pierpoint Langley, the rival of the Wright brothers, failed to launch this aircraft, his *Aerodrome A*, from the roof of a houseboat.

Unfortunately, Kress never managed to fly his frail aircraft because it capsized while taxiing on the Tullnerbach Reservoir and then sank. Although the Kress was the first full-sized petrol-powered aeroplane to be built, the Kress floatplane was overcome by the not inconsiderable weight of its engine, doubtless primitive float design and a lack of ballast for stability.

Neither Langley's nor Kress's work should be viewed as opportunities lost. Both reflected the chauffeur concept of flying an aeroplane rather as one might drive a car along a road. There was no provision for control in roll and had either design managed to take off they would have failed to fly safely because of this. Any attempt to turn would have meant a loss of control.

Shortly after Langley had made his unsuccessful attempts at flight, the problem was solved by a landplane. Several years of increasingly successful experiments with biplane gliders, followed by the development of their own lightweight power-plant, resulted in the first tentative aeroplane flights by the brothers Wilbur and Orville Wright. True flight, as opposed to powered leaps or hops in which the force that counts is the aircraft's take-off run and the resulting momentum, requires that the engine is able to sustain the aircraft's progress through the air, and the flight must be long enough for control to be exercised by the pilot, while the aircraft must land on a spot no lower than that from which it took off. Such requirements were essential during the early days to distinguish between aeroplane flight and the efforts of such unsuccessful pioneers as Mozhaiski, du Temple and Ader. The toss of a coin gave Wilbur the opportunity of making the first flight on 14th December, 1903, but he pulled the aircraft, later known as the Wright Flyer I, off the take-off rail too sharply and repairs were necessary before an attempt could be made by Orville on the morning of 17th December, 1903, followed by three further flights that morning with the brothers taking turns at the controls. Some of the Wrights' rivals, including the French pioneer Gabriel Voisin, dismissed the Wrights' efforts asserting that take-off was achieved by use of a launching apparatus, but this did not come until much later and meanwhile, in 1904 and 1905, the Wrights refined their techniques with successive aircraft, culminating in the first practical aeroplane, the Wright Flyer III, of 1905.

Unfortunately, denied attention and recognition, not only by the Europeans who refused to accept that they had been beaten into the air by Americans but also by the United States Army, the Wrights suspended further trials until 1908, when they conducted a conclusive series of demonstration flights in Europe and the United States which also showed the inadequacy of European design at the time. It was much later before Wright aircraft were to appear with floats and that was to be the result of other pioneers adapting the brothers' standard designs.

Nevertheless, this was not a completely arid period for aviation in Europe and the hydro-aeroplane, in particular, received considerable attention from the European pioneers. Gabriel Voisin – working at Billancourt at what he and his brother, Charles, were later to proclaim as the world's first aircraft factory – designed and built two float-gliders in conjunction with his customers, Ernest Archdeacon and Louis Blériot. At first Voisin tended to work to the designs of his customers rather than his own. On 8th June, 1905, the Archdeacon aircraft, or Voisin–Archdeacon as it is sometimes called, was towed off the River Seine near Paris by an Antoinette-powered motor-launch, with Gabriel Voisin himself seated at the controls for a glide of about 500 feet. This flight was repeated ten days later, on 18th June, the same day that the Voisin–Blériot float-glider made its first flight, again with Voisin at the controls of these aircraft, although the Blériot model collapsed, giving Voisin a soaking. Even the Voisin–Archdeacon made some ungainly landings, perhaps accounting for the ten-day interval between flights.

Both the Voisin gliders for Archdeacon and Blériot incorporated a combination of the Wright glider and the Hargrave box-kite features.

Problems with the Voisin–Archdeacon float-glider during tests on the River Seine during June 1905.

The Voisin–Blériot float-glider prepares for take-off, also during June 1905. The Hargrave box-kite features are particularly prominent on the tailplane.

The Wright influence manifested itself in a forward elevator, while the biplane mainplane, although the Blériot aircraft was in fact a sesquiplane with differing spans to the upper and lower wings, used Hargrave-type box-kite cells, as did the biplane tailplane. This type of aircraft was to be the basis of early European flight, but neither machine incorporated control in roll, since the Wright concept of wing-warping was little understood and its importance was not appreciated in Europe.

More trials followed on both gliders. The Archdeacon model was taken to Lake Geneva for further tests, which were unsatisfactory, and was then sold to a Frenchman named Bellamy. He took it to England in 1906, added an engine and two propellers and made an unsuccessful attempt at powered flight, although it is not known whether this was with the original floats or whether the machine had been modified as a landplane. The Blériot, which differed from the Archdeacon not only in being a sesquiplane with a smaller span to the lower wing but also in having a dihedral angle to the mainplane and a pronounced camber to all lifting surfaces, was rebuilt as a powered biplane with a 24-hp Antoinette engine but remained a failure: both as a floatplane on Lac d'Enghien, first with one and then with two 24-hp Antoinette engines, and then when, fitted with wheels, it crashed during its take-off run, never becoming airborne. However, scant though it was, there was enough of the smell of success in these designs to provide further encouragement for the Voisin brothers.

Slightly more successful was the machine built at Billancourt by the Voisin brothers for their client, Léon Delagrange, in 1907. Using a similar configuration to the Voisin–Archdeacon, although without the mainplane side panels, the Voisin–Delagrange I used a 50-hp Antoinette engine and as a landplane actually succeeded in making a few tentative hops during March, 1907. Subsequently, floats were fitted for trials on Lac d'Enghien, but these were unsuccessful and the wheels were soon replaced, although the aircraft crashed during a later series of tests. A development of this machine, the Voisin–Delagrange II, flew successfully in 1908 and enjoyed the distinction of carrying the first woman passenger, but this was solely as a landplane. In fact, two Voisin–Delagrange IIs were built, the second being for a private customer, but it is not certain that the second aircraft ever took to the air.

Nevertheless, in spite of the problems and the rivalry with the Americans, it was to a Frenchman that the distinction of making the first seaplane flight eventually fell.

After qualifying as an engineer in 1905, Henri Fabré, eldest son of a wealthy shipping family and a native of Marseilles, started to study the work of Blériot, Ferber and the Voisin brothers, before experimenting with a design of his own in 1909. This aircraft, a monoplane with three Anzani engines driving a single propeller, was built but never flown. Nevertheless, Fabré almost immediately set about the design and construction of his second aircraft, which he called the *Hydravion*, a novel design which incorporated a number of distinctly nautical features, including canvas-covered wings on which the canvas could be reefed up to the spars when the aircraft was being stored. Rather more practical were a number of Wright design features, including wing-warping for control in roll and a tail-first biplane layout. Wing-warping was the slightly impractical 'helical twisting of the wings', as the Wrights described it, which was superseded by ailerons for control in roll as aircraft became larger and faster. Unlike the twin-float Voisins, the Fabré aircraft used an unusual three-float tricycle undercarriage, while power was provided by a single 50-hp 7-cylinder Gnome rotary engine, which drove a pusher propeller.

On 28th March, 1910, having never flown before, even as a passenger, Fabré taxied out in the *Hydravion* at La Mede, near Marseilles, and attempted to take off, but even at 30 mph the aircraft failed to unstick. Clearly filled with his share of the optimism with which so many of the pioneers were blessed, Fabré tried again later that same day and on this attempt the frail machine managed to fly for 1,640 feet at six or seven feet above the surface of the water. Other flights followed and on his fifth flight, which was on 29th March, Fabré managed to remain airborne for almost four miles, while reaching an altitude of more than sixty feet. Unfortunately, the frail aircraft was seriously damaged on 18th May, but it was rebuilt in time to be shown at the Salon de l'Aéronautique in Paris during October, 1910. Later development work on the *Hydravion* included the addition of retractable yacht-style keels on the floats, and although these were unsuccessful on the front float, the keels improved the directional stability during water running on the rear two floats.

At Monaco in 1911, the *Hydravion* made two flights with Jean Becue at the controls, although on the second flight the aircraft broke up while landing in the surf close to the beach. While Fabré is believed to have attempted to build a number of developments of the *Hydravion* after this, it is not known whether any were completed before he too was compelled to abandon experiments due to a shortage of funds. Fabré retained a close connection with the hydro-aeroplane, or at least with floatplanes or seaplanes, after this, building floats for many pre-World War I French manufacturers, including Voisin, and indeed, at the 1913 Monaco floatplane meeting, all of the aircraft used Fabré floats! Although undoubtedly the first floatplane or seaplane, the Fabré could not be considered to be a truly practical machine, and this too had to come from the other side of the Atlantic.

A motor-cycle manufacturer – the Wrights had been bicycle manufacturers and the Voisin brothers motor-car dealers – Glenn Hammond Curtiss first became interested in aviation when, with Doctor Graham Bell, he was one of the founders of the Aerial Experimental Association (AEA) at Hammondsport, New Jersey, in 1908. From the start, the AEA's designers produced sound workable aircraft, and it was their second design, the *White Wing*, that Curtiss was to fly for the first time, making five flights on 21st May, 1908. Curtiss's own first aircraft design was also for the AEA, the *June Bug*, which actually did fly during June, 1908, and went on to win the *Scientific American* magazine trophy for the first officially observed flight in the United States. It made a total of fifty-four flights, the last of which, on 4th July, covered 5,090 feet in just under two minutes.

It was as a result of his success in the *June Bug* that a long and, for the Wrights, damaging fight began between Curtiss and the Wright brothers over what the brothers regarded as an infringement of their patents, especially those covering wing-warping, even though the AEA aircraft and those of major European pioneers such as Blériot and Levavasseur, the owner of the Antoinette concern, used ailerons, which were to be the most practical solution to the problems of control in roll.

Still during 1908, in December, Curtiss and the AEA made their first attempt to fly a hydro-aeroplane, called the *Loon*, but without success. Again, in 1909, Curtiss attempted to build a floatplane by fitting a landplane, probably the *June Bug*, with a canoe as a float, but still without success. Further experiments with the *Loon* followed during 1910, although the design still did not meet Curtiss's expectations of it. This was undoubtedly due to poor float design.

The pioneers were suffering from a problem which was to affect seaplane development for many years, the need for floats, or even flying-boat hulls, which offered stability on the water and which would allow the aircraft to take off, or 'unstick'.

Curtiss nevertheless continued his success with the *June Bug* with his *Golden Flier*, another landplane, which set the earliest aeronautical records at the Rheims Aviation Meeting in 1909. As a result, this aircraft was often referred to as the Curtiss *Rheims Machine*. However, Curtiss had not forgotten the floatplane and on 26th January, 1911, he successfully tested the world's first practical floatplane, a modified

Golden Flier with a single main float and two smaller wingtip floats in place of the wheeled undercarriage, at San Diego, California, causing a contemporary to remark that 'he had made the first significant advance in aviation since the Wright brothers'. On 25th February, Curtiss took up the first passenger to be carried in a seaplane, using this same aircraft, while in that same month he modified the aircraft by fitting wheels to the floats to produce the first crude, but workable, amphibian. Recognition was not long in coming, with Curtiss being awarded the Collier Trophy for the most significant advance in aviation for his work on hydro-aeroplanes.

Very much the new toy, the Curtiss floatplane opened up a whole new concept in aeronautics and became the object of an exhausting series of trials and experiments. Completely unscheduled and not at all like a toy, the aeroplane made the first recorded air-sea rescue, picking up a ditched airman from Lake Michigan in August, 1911, when Hugh Robinson landed to pick up the pilot of a landplane who had made an emergency landing in the lake. On 7th September, Lieutenant T. G. Ellyson, USN, took off in a Curtiss floatplane from an inclined wire, the first launch of a workable hydro-aeroplane, while another modified Curtiss *Golden Flier* landplane featured in the early experimental flights from US warships at the hands of another young United States Navy pilot, Lieutenant Eugene Ely.

Of course, Curtiss could not have it all his own way, there was far too much interest in aviation by this time, and development was still within the financial reach of individuals before the rising cost of technology put this first in the hands of companies and then of governments. On 3rd August, 1911, a report in *The Times* recorded an event of the previous day: 'The airman Colliex flew from Issy-les-Moulineaux this morning in a biplane fitted with floats and descended on the Seine, where he stopped on the water and floated. After another flight from the land and descent upon the water, the aeroplane was taken in tow and brought to Issy'. The term, 'flight from the land' probably means the shore, since the aeroplane, a standard Voisin machine fitted with Fabré floats, was not an amphibian.

A Royal Navy officer, Lieutenant Oliver Schwann, bought an aeroplane at his own expense, probably a Wright biplane built by the brothers' British licensees, the Short brothers, and fitted this with floats and gas-bags. He was able to make a water take-off in this machine in November, 1911. A fellow officer, Lieutenant Arthur Longmore, RN, later to become Air Chief Marshal Sir Arthur Longmore, RAF, flew a Short S.27 biplane fitted with air-bags onto the River Medway on 1st December, 1911, and subsequently

The most consistently successful of the hydro-aeroplane pioneers was Glenn Curtiss, although he is sitting at the controls of his *Rheims Machine* landplane in this photograph of 1909.

Unsuccessful, the *Loon* was the first attempt by Curtiss at building a floatplane, working with his associates in the AEA in 1908.

First flight from a British warship was by Charles Rumney Samson from HMS *Africa* in this Short S.27 biplane fitted with air-bags, in January, 1912.

For the 1912 Naval Review, the battleship HMS *Hibernia* was fitted with a launching ramp for a Short S.27 biplane.

succeeded in taking off again. Essentially a Wright aeroplane design, the S.27 owed much to the influence of the Wright A biplane.

The Short S.27 had also been used for the Royal Navy's early experiments in shipboard take-offs, with Lieutenant Charles Rumney Samson flying one of these aircraft from the battleship HMS *Africa* while she lay off Sheerness on 10th January, 1912. Although the aircraft could not be described as a fully fledged seaplane on this occasion, it was nevertheless equipped with air-bags in case of an emergency landing on the River Medway, a safety precaution which reputedly 'heartened' Samson's admiral. Samson had already attempted a water take-off in this aircraft on 28th November, 1911, but had crashed on landing.

The United States Navy ordered a Curtiss floatplane in July, 1911, giving it the designation A-1, although this aeroplane was later to become known as the *Triad* with the addition of wheels to the floats, indicating the ability of the machine to operate in the air, or on land or sea! Of rather more practical benefit at this time was the fact that Curtiss had introduced a 'step' to the float of the *Triad*, making take-off easier as the aircraft would unstick more readily than any of its contemporaries. Other features of the *Triad* included the setting of an altitude record for hydro-aeroplanes on 21st July, 1912, by reaching a height of 900 feet. Eventually, twelve of these aircraft entered USN service, being redesignated Airplane Hydro (AH), while the United States Army Signal Corps introduced three 60-hp Curtiss Type E biplanes, fitting one of these with floats. Meanwhile, it was a Curtiss AH of the USN which made the first successful flight from the compressed air catapult invented by Captain Irvin Chambers, USN, when Lieutenant T. G. Ellyson flew an AH off a barge which had been fitted with the catapult, at the Washington Navy Yard on 12th November, 1912.

Curtiss then turned his growing talent to the flying boat, initially developing one of his seaplanes into a rudimentary flying boat with the single float enlarged to provide a hull, and this two-seat machine first flew from San Diego on 10th January, 1912. Further work on the flying boat theme enabled Curtiss to produce his F-boat later that year with a single 100-hp Curtiss OXX 8-cylinder 'V'-type engine mounted between the wings of the biplane, or 'inter-wing', driving a pusher propeller. Capable of a cruising speed of about 70 mph, the F-boat enjoyed the distinction of becoming the first aircraft to fly with the new Sperry gyroscope auto-pilot, which was first demonstrated on this machine by Lawrence Sperry, flying from Lake Keuka in New York State. A British manufacturer, White and Thompson, obtained a licence from Curtiss to build the F-boat, using hulls built by the boat-building firm of S. E. Saunders at Cowes on the Isle of Wight. Before long these boat-builders were to become one of the world's leading producers of flying boats in their own right, although they also produced hulls for other manufacturers from time to time. Both the United States Navy and the United States Army Air Corps were to order the F-boat in considerable numbers, although the USN's initial order was for just five of these aircraft, designated the AB, early in 1914.

The Curtiss F-boat is generally believed to have been inspired by the work of a French engineer and pilot, F. Denhout, who designed and built a 50-hp Gnome-powered biplane flying boat with the engine mounted between the wings driving a pusher propeller and with the tailplane mounted on twin booms, which he flew off the Seine at Juvisy on 15th March, 1912. Unfortunately, the aircraft crash-landed and needed to be rebuilt, although after rebuilding with some modifications it was back in the air again on 13th April, 1912. Denhout's work attracted the attention of his fellow countrymen, Donnet and Lévèque, who formed a company to build and market the new aircraft. Denhout, not surprisingly, became a director of the new firm, Donnet–Lévèque, which acquired premises at Quai de Seine, Argenteuil. Another director was a French naval officer, Lieutenant de Vaisseau Jean Conneau, who for professional reasons had to fly in public under the pseudonym of André Beaumont. While the new company found its feet, as it were, the first two production Donnet–Lévèque Type As, as the aircraft were now designated, were built at Juvisy and flown to Argenteuil on 26th July, 1912, by Beaumont, alias Conneau. On 9th August, Conneau attempted a Paris–London flight with a Type A, although this had to be abandoned after an accident while taking off in Boulogne Harbour, one of the staging and refuelling points along the route. Nevertheless, in September, while flying at a display organized by the Aéro Club de Belgique, Conneau won the King of Belgium's prize. Orders for the aircraft followed after this, with one for the Royal Navy, using an 80-hp Gnome engine, delivered to Eastchurch on the Isle of Sheppey on 20th December; three others were built for Austria-Hungary; two for Denmark; and a single aircraft for Sweden. The Type A was also exhibited at the Paris Aéro Salon in October, 1912, incorporating an amphibian undercarriage, although, in spite of its success as a flying boat, the aircraft could not be regarded as a successful amphibian.

The accolade of being the first practical amphibian in fact belongs to the oddly named Sopwith *Bat Boat*, designed by Tommy Sopwith with the intention of winning the £500 Mortimer Singer prize for the first British amphibian. Again, S. E. Saunders built the hull, which was very much speedboat-shaped and only 21 feet long. On the early trials the biplane used a single 90-hp Austro-Daimler 4-cylinder engine in the inter-wing position, while twin booms supported the monoplane tailplane. After being shown in public at the Olympia Aero Show in February, 1913, the aircraft was successfully flown at Cowes, but in order to qualify for the Mortimer Singer prize, a British engine had to be substituted for the Austro-Daimler engine, and so a 100-hp Green engine was fitted, with the result that the *Bat Boat* was able to win the prize on 9th July, 1913. The Royal Navy bought the *Bat Boat* after it had been converted from an amphibian to a flying boat, and after it crashed in Royal Navy service a replacement was ordered. An enlarged version was built as the *Bat Boat 2* and displayed at Olympia in March, 1914, with a 200-hp engine, and Howard Pixton planned to fly this aircraft in the Circuit of Britain air race, but World War I intervened and the race was cancelled.

Meanwhile, all had not been going well for the Wright brothers, who had dissipated much of their energy in an unfortunate and ultimately fruitless lawsuit intended to protect their patents. Their main opponent was Glenn Curtiss, although other pioneers throughout the world also waited apprehensively. The Wright technique of wing-warping was little used by this time, and the death of Wilbur from typhoid on 30th May, 1912, led eventually to Orville losing interest in the company he and his brother had founded. This, combined with the onset of World War I, brought the debate and legal action over the Wright patents to an end. However, the struggle had reached such a pitch at first that in 1914 Curtiss still felt the need to modify Langley's *Aerodrome A*, flying this aircraft in an altogether not very praiseworthy attempt to discredit the Wrights' claims.

Technically, the Wright brothers had already been surpassed, and such activity as did occur owed much to their students and other designers who passed through the company. Early in 1912, one Wright student, Frank T. Coffyn, equipped a Wright B biplane with aluminium floats to produce a hydro-aeroplane variant, flying this successfully from the ice-covered Hudson River, taking off from a spot close to the southern end of Manhattan, circling the Statue of Liberty and then returning to make a safe landing. In 1913, the Wright Company produced the Model G Aerobat, designed by Grover C. Loening, who combined the Wright aerodynamic surfaces to a flying boat hull, while another success came from a short association between the Wright concern and William Burgess.

The Burgess *Military Tractor* or *Coast Defense Hydro*, which entered US Army service in 1913.

The owner of a shipyard at Marblehead, Massachusetts, Burgess had switched to building aircraft in partnership with Augustus Moore Herring, who had enjoyed the unparalleled distinction of having worked with Samuel Langley at the turn of the century, then with the Wright brothers' guide and mentor, Octave Chanute, and then with Curtiss, having formed a short-lived Herring and Curtiss Company before joining Burgess. Burgess himself had already savoured the joys and disappointments of early success in aeronautics: in partnership with Harold Gill in May, 1912, he had built a seaplane hoping to win the $15,000 Edwin Gould Prize for the first twin-engined hydro-aeroplane capable of meeting the now accepted standard of remaining airborne on the power of only one engine. The twin-engine, four propeller, Burgess–Gill was indeed capable of this achievement, but when it became apparent that it was to be the only aircraft to take part in the competition, the sponsor decided not to award the prize! Gill lost interest, but Burgess obtained a Wright licence and produced the Burgess–Wright H biplane seaplane, otherwise known as the Burgess *Military Tractor* or the Burgess–Wright Model 1, before passing to the United States Army Air Corps, who named it the *Coast Defense Hydro*.

Meanwhile, back in Britain, the Royal Navy had become accustomed to water take-offs and landings. A small selection of Deperdussin, Nieuport and Short aircraft were taken to the 1912 Naval Review, where the battleship HMS *Hibernia* was fitted with a ramp over the forecastle for launching aircraft and became known, not quite correctly, as a hydro-aeroplane mother vessel! However, one of the four aircraft at the Review, which was held in Weymouth Bay, was fitted with floats. This was a Short S.27 biplane, named by her pilot, the ebullient Charles Rumney Samson, as HMS *Amphibian*, happily ignoring the fact that the torpedo-shaped floats contained no provision for amphibious landing gear! Nevertheless, in one of many flights at the review, Samson carried a message for King George V, landing his S.27 alongside the royal yacht, *Victoria and Albert*, so that the messenger could be carried by dinghy from aircraft to ship. Shortly afterwards, the Royal Navy opened the first of a chain of North Sea seaplane stations on the Isle of Grain intended to help to defend Britain's east coast, while the French came to convert the first warship to become a permanent seaplane-carrier, the cruiser *La Foudre*, which was recommissioned in her new role in late 1912. The Americans also opened their first seaplane station at this time, at Guantanamo in Cuba, but closed it again early in 1913 after it had been open for just a few months.

Shorts, meanwhile, were moving away from the Wright pattern of aeroplane to the more conventional and up-to-date tractor biplane, doubtless encouraged by the prospect of orders from the Royal Navy and the keen interest which was being taken in naval aviation by the then First Lord of the Admiralty, Winston Churchill, who on at least one occasion allowed Samson to act as his pilot for a flight in one of the new Short S.41 biplane seaplanes, aircraft of what would soon prove to be the classic British pattern, with a single engine at the front of the fuselage driving a single tractor propeller, twin main floats and smaller stabilizing wingtip floats. Not all S.41 flights enjoyed the stamp of official approval, however, and when one Frank McClean flew through Tower Bridge and up the River Thames in his S.41 floatplane to land at Westminster, the public acclaim for this achievement was also marked by strong police disapproval and a reprimand, although he does not seem to have been prosecuted. Short's next seaplane, the S.45, was less conventional than the S.41, at least for a British aircraft, having the American feature of a large single float, although in fact this aircraft was more usually seen in landplane form.

The spread of the seaplane was in fact almost unstoppable by this time, and indeed a special seaplane event was added to the Monaco powerboat meeting in March, 1912, with no less than seven seaplanes competing, including one or two would-be amphibians, and the race was won by Fischer flying a float-equipped version of the latest Henri Farman biplane. It was after the 1912 Gordon Bennett race for landplanes that Jacques Schneider announced his intention to award a trophy for a contest for hydro-aeroplanes, either seaplanes or flying boats, which he believed to be potentially far superior to landplanes and offered far greater opportunity for raising air speeds. For a time, as

Official support for the Royal Navy's aviators came from Winston Churchill, First Lord of the Admiralty, about to take off here in a Short S.42 with Samson as pilot.

Shorts' next aircraft, the S.45, had a single float, a rarity on British seaplane designs.

An early flight from a French warship: a Caudron G.3 leaves the cruiser *La Foudre*.

Glenn Curtiss flies the much modified Langley *Aerodrome A* in 1914, in his bid to discredit the Wright brothers' patents.

we will see in Chapter 3, events were to prove him right!

The threat of war was becoming more serious in Europe, however, and the leisurely peacetime development of the flying boat and seaplane was shortly to be brought to an end, while already both types of hydro-aeroplane were being seen as war planes with definite tasks in any conflict. However, for a while, the Americans were able to continue to develop aviation in much the same haphazard but enjoyable and exciting way to which the early pioneers had been used to, while Europe readied itself for battle.

Indeed, the Americans even managed to introduce the world's heavier-than-air passenger service using small flying boats. A little-known aircraft company, the Benoist Aircraft Company of St Louis, Missouri, was the builder of a small two-seat flying boat with a single 75-hp Roberts engine driving a single inter-wing propeller. Benoist also ran a flying school, a not unusual sideline for manufacturers during the early days of aviation, and the chief instructor, Tony Jannus, enjoyed the distinction of taking Albert Berry up to a height of 1,500 feet for a series of tests with a parachute – the first such tests in the United States. Clearly a man with other interests than flying lessons, Tony Jannus became the first pilot of the new St Petersburg and Tampa Airboat Line when it was founded on 1st January, 1914. This operated a couple of Benoist Type XIV flying boats across a bay in Florida for a fare of $5 single, providing that the one and only passenger on each flight did not weigh more than 200 lbs! The journey around the bay between Tampa and St Petersburg was 36 miles by road, while the flying boat, cutting across the bay, flew just 20 miles, taking 20 minutes against two hours by road. The airline operated a maximum of just two flights daily, and in fact the operation was really more of a scheduled or perhaps semi-scheduled air taxi operation. It lasted for just a few months.

However, inspired by the success of the flying boat, Benoist were to develop a six-seat flying boat, the Model C with two 100-hp Roberts engines, which first flew in 1915. Unfortunately, war in Europe and the risk of American involvement meant that there were no prospective airlines ready for this aircraft at that time, and by the return of peace, in 1918, Benoist no longer existed.

Indeed, despite the early start of the Benoist flying boats on their short route within Florida and of the early German Zeppelin airship services which operated between 1910 and 1914 within Germany, it was not to be until 1919 that regular airline services were to appear once again, and at first these were to be operated exclusively by landplanes.

2
WORLD WAR I

In a violent world, it was inevitable that the hydro-aeroplane in whatever form, be it seaplane, or floatplane as these were known at first, or flying boat, would soon be adopted for military and naval use. Despite the claim by the Wright brothers that they had believed that their invention was a machine which would make further wars impossible – perhaps this was a reference to the deterrent effect of aerial observation rather than anything else, since the Wrights were well to the fore in understanding and attempting to make the most of the aeroplane's military potential – the aeroplane added yet another dimension to warfare. Thanks to the aeroplane, the city dweller hundreds of miles away from the battlefront or the coast found himself on the front line and subject to ever more devastating aerial attack. One shouldn't become too sanctimonious about this, aeroplanes themselves are not wagers of war, they are either defensive or aggressive depending on which side one stands, since there are really no such things as defensive or offensive aeroplanes: nice though it may sound, a nation cannot defend itself with fighter aircraft alone and sooner or later the real defence is found to lie in an attack on enemy airfields and factories.

Even before the practical aeroplane appeared, military applications had been sought for kites, balloons and airships, and from 1908 onwards, the aeroplane was fitted with guns, wireless telegraphy equipment, bombs and grenades in an attempt to assess its possibilities. Certainly, most people, even in the military, saw the aeroplane primarily as a reconnaissance machine, but there were those who foresaw a more aggressive rôle and the idea of aerial bombardment from flying machines does in fact date from the Middle Ages! For the maritime role of the aeroplane, regardless of whether it was a landplane or seaplane, a significant step forward came in July, 1912, when Rear-Admiral Bradley A. Fiske patented a method of air launching a torpedo. An early trial with an air-launched torpedo came on 28th July, 1914, on the opposite side of the Atlantic, when Lieutenant Arthur Longmore, RN, made the first drop of a standard naval torpedo from an aeroplane while flying at Calshot on Southampton Water in a Short Folder biplane which carried the torpedo on a rack between the floats. As its name suggests, the Short Folder was a

The Royal Navy entered World War I with a single seaplane-carrier, the converted collier HMS *Ark Royal*.

machine with folding wings, developed from the earlier S.41 and S.45 for operations from seaplane-carriers. Another torpedo-carrying aircraft at this early date was the French Borel seaplane, a monoplane designed and introduced in 1914.

Indeed, by 1914, the seaplane-carrier was a fact of naval life. The French Navy, the Marine Nationale, had already adapted the cruiser *La Foudre* as a seaplane-carrier operating a variety of aircraft apart from the Borel and including the Caudron G.3 biplane. In 1913, the Royal Navy had converted the light cruiser HMS *Hermes* to act as a carrier for the new Short Folder biplane at that year's Naval Review, but this ship operated in her new rôle for only a short period before being converted back to a light cruiser – a much-needed class of vessel with World War I looming – and a collier was converted to take her place and named HMS *Ark Royal*, a not inconsiderable promotion for a vessel of such humble origins! In Japan, the Imperial Japanese Navy followed events in Europe closely, converting a naval transport, the *Wakamiya*, to operate as a seaplane-carrier, operating a mixture of French and Japanese aircraft, including Farman, Blériot, Otari and Ushioku seaplanes. The Imperial Russian Navy had acquired some sixty flying boats and seaplanes of various types by the outset of World War I, while the United States Navy could only boast a small number of Curtiss AH Triad and Wright seaplanes, and some Curtiss and Burgess–Dunne flying boats. In fact, the USN was in danger at this time of seriously underestimating the value of the aeroplane, with a report by the chief of Naval Ordinance suggesting that even the 600-lbs lift of the more powerful aircraft would be inadequate to carry a torpedo capable of sinking an enemy warship.

Of these aircraft, the most unusual and interesting was the Burgess–Dunne. Before the outbreak of war in Europe, William Burgess had severed his connections with the Wright Company, leaving the United States to work with the unconventional English designer and army officer, Captain John W. Dunne, who had been responsible for a series of novel swept-wing tail-less biplanes which had been flown with some considerable success. Working together and applying Burgess's seaplane experience to Dunne's own not inconsiderable talents, the series of aircraft known as the Burgess–Dunne were built with a large, single, central float and small stabilizing wingtip floats. The first Burgess–Dunne was ready for trials in January, 1914, but did not enjoy immediate success and the machine was damaged during an attempt at take-off. After rebuilding, during which the opportunity was taken to convert the aircraft from a single-seat to a twin-seat configuration, the first Burgess–Dunne made a successful first flight in March, 1914. After completing its trials, two aircraft were ordered by the United States Navy, while both the United States Army Signal Corps and the United States Coast Guard ordered single examples of the aircraft, all of which bore the USN designation of AH–7. The first AH–7 was a standard Burgess–Dunne machine with a single 135-hp tail-mounted Salmson M9 engine driving a pusher propeller, and with a maximum speed of around 75 mph. However, later USN orders specified 140-hp Sturtevant engines and a stretched fuselage nacelle and certain airframe modifications, but the type did not take well to these changes and, designated the A–54, it proved to be difficult to control in a dive. Other Burgess–Dunne aircraft followed, some of which had uniform wing camber rather than the variable camber originally favoured by Dunne, but by the early years of the war Dunne was elsewhere employed and there was less time for experimental designs, leaving Burgess to return to the United States during the winter of 1915–16 to join the Curtiss Aeroplane and Motor Company.

Another futuristic, but nevertheless more conventional, British design at this time was the Pemberton-Billing P.B.1, the predecessor of the Supermarine line of aircraft which were eventually, in the period between the wars, to raise the seaplane to a peak of excellence. The name Supermarine itself started as Noel Pemberton-Billing's telegraphic address when he opened his seaplane and flying boat factory at Woolston, on the outskirts of Southampton, before the onset of World War I. The P.B.1 was his first design, a remarkably streamlined flying boat with a biplane mainplane and a monoplane tail, with a single 50-hp Gnome rotary engine mounted above the hull and in front of the mainplane. In spite of its small size and even smaller power output, the P.B.1, just 27 feet long and with a 30-foot wingspan, could manage 50 mph, although this was far less than the streamlining seemed to promise. However, the sleek lines of the aircraft caused it to be one of the star exhibits at the Olympia Aero Show in March, 1914, although at this time the aircraft had still to prove itself in the air. The early trials were none too successful, so the position of the cockpit, which was originally aft of the mainplane, was moved forward and the position of the engine was also adjusted. Yet, in spite of these changes, the aircraft crashed during its test programme. A number of impractical designs followed, including the P.B.3 and P.B.5 flying boats, and another design with detachable wings so that the aircraft, probably intended for the private owner, could convert to a cabin cruiser. A number of other designs were not built after Noel Pemberton-Billing left his company to join the Royal Naval Air Service, with the one exception being the P.B.25 Scout flying boat, although this aircraft did not enter production. The brilliance and later glory of Supermarine owed much to the efforts of Hubert Scott-Paine, who assumed control after Pemberton-Billing's departure, and of course later to the genius of Reginald Mitchell. However, during World War I, Supermarine was no more than a builder of other manufacturers' designs, under licence.

The homeland of the Wright brothers was falling behind in aeronautical development at this time, while the main progress was being made in the United Kingdom and Europe. Development of new American designs was being hampered by a lack of suitable powerplants, and this was to be a weakness throughout the war years. Several hundred Hispano-Suiza engines had to be acquired from France for American airframe manufacturers, and a quantity of these was allocated to Wright–Martin for flying boat construction. However, the enterprising American motor industry quickly came to grips with the problem, designing, developing and putting into production an American aero-engine, the Liberty. Even so, by the time the United States entered the war on the side of the Allies – Britain, France and Italy – operational aircraft were still having to be acquired from abroad, with purchases including a number of Caproni flying boats. The United States Navy's own Naval Aircraft Factory, which had been opened at the Philadelphia Navy Yard, also built some US-designed aircraft, the Curtiss H–16 flying boat, but the work of Curtiss, and the mobilization of other manufacturers to support Curtiss's own efforts, was still not enough to meet the demands for aircraft from the United States armed forces. Indeed, even Europe could not meet American demand for aircraft, with the French, for example, only able to deliver one-third of the aircraft ordered by the USN before the cessation of hostilities in 1918 made such strenuous production efforts unnecessary!

As more aircraft became available and American involvement in World War I became more certain, the United States Navy and United States Marine Corps (USMC) did not seek to emulate the example of the British Royal Naval Air Service, which had a considerable number of its squadrons allocated to home defence fighter and front line duties, but instead, from the outset the Americans concentrated on maritime reconnaissance and anti-submarine warfare. In the last year of the war, the USMC actually posted a squadron of eighteen Curtiss seaplanes and flying boats to the Azores, giving them a commanding position over a substantial area of the mid-Atlantic. However, there were some bombing missions, with a definite nautical flavour of course, including the raid by USN Caproni biplane flying boats on the Ostend U-boat pens on 15th August, 1918, shortly before the war ended,when 1,250 lbs of bombs were dropped.

Maritime reconnaissance became a vitally important element in a conflict which saw submarine warfare progress at least as rapidly as aerial warfare. Allied merchant shipping suffered terribly, with new tonnage not exceeding losses until 1917, and even in that year the Allies lost 2,600 ships to a German U-boat fleet of just 127 craft. Devoid of sonar equipment, surface vessels were dependent on the help which could be provided by escort vessels and by the still poorly equipped Curtiss, Caproni, Felixstowe and Supermarine flying boats, including Supermarine-built Felixstowe flying boats. The flying boat offered the best means of maintaining long-range maritime reconnaissance patrols, but such aircraft could never pretend to be fast, while their range and payload meant that by modern standards the cover they provided tended to be little more than coastal. Flight safety, the vulnerability of the aircraft to anti-aircraft fire or enemy fighters and the effects of weather and climate on the then current standards of construction, all meant that the casualty rate was very high, and so too was the rate at which aircraft had to be retired from operational service. By the Armistice, the Royal Air Force, which succeeded both the RNAS and the Royal Flying Corps in April, 1918, was operating only three out of fifty Curtiss H.4s built for the RNAS, although the new service was operating more than fifty Felixstowe F.2s, almost a hundred F.3s and a growing number of the popular F.5s. To help flying boats to survive, the RNAS introduced a dazzle camouflage finish which was supposed to make it more difficult for German fighters to spot the aircraft, in 1918.

One of the main problems in keeping flying boats serviceable at this time was the tendency for the wooden hulls to become waterlogged or to spring leaks, affecting buoyancy and the take-off weight during the life of the aircraft. The lighter weight hull of the flying boat compared to that of a wooden ship aggravated this situation, so that while many of the old 'wooden wall' sailing ships built a hundred years or more before the flying boat was even heard of enjoyed a long active life, the aeroplane, with its constant, forceful splashing down into the sea and high speeds on the take-off run, was more vulnerable.

Not all of the flying boats were confined to Europe. The Allies themselves included Japan amongst their number, with the Imperial Japanese Navy using its small number of flying boats to enforce a blockade of isolated Germany treaty ports on the mainland of China. Another ally was Brazil, which had received its first seaplane, a French Bossi monoplane, at the seaplane station established by the Brazilian Navy near Rio de Janeiro in 1913. Two more Bossi seaplanes followed within a year, and by 1916 the seaplane station had grown into a fully fledged naval air station at Ilha de Enxades in Rio de Janeiro Bay. Brazilian pilots were trained in Britain as well as in their own country, and experienced airmen were seconded to the Italian armed forces. Portugal also supported the Allies, but only army units saw active service, while the Aviação Maritima, founded in 1917 and with air stations at Lisbon and Aveiro, sent its pilots to Britain, the United States and France for training and received its first aircraft, Fairey Campania floatplanes, too late for service in World War I. A number of Felixstowe F.3 flying boats were also delivered post-war after refurbishment by Fairey.

Felixstowe F.2 flying boats in camouflage, the first of a successful series used by many leading air arms.

The neutral countries also had to struggle to obtain aircraft for their still young air arms. Sweden had only four aircraft at the outbreak of war and only two of these belonged to the Royal Swedish Navy, but a further twenty-five aircraft were delivered during the war years after the formation of a Swedish aircraft industry, the Thulins Aeroplanfabrik. Naval aviation in Denmark struggled through the war, operating licence-built flying boats from a base close to Copenhagen. The Netherlands Marine Luchtvaartdienst, or Naval Air Service, was formed on 18th August, 1917, with bases at De Kroog and De Moh, and operated a variety of aircraft, many of which were German or Allied types which had been interned after force-landing on Dutch territory, although a number of Swedish Thulins aircraft were also acquired.

However, the real progress tended to lie in the development of the British, French and American air arms. This was frequently dramatic. The United States Navy possessed just a handful of seaplanes in 1915, yet by the Armistice of 1918, it had some 1,400 flying boats and seaplanes, with another 624 aircraft in the United States Marine Corps, while the French Navy's Service Aéronautique had 1,264 aircraft at the Armistice, including FBA and Tellier flying boats. The Royal Navy, after the merger of the RNAS and the RFC, had none, with all service aviation centred upon the new Royal Air Force, the world's first autonomous air service, free of army or navy control.

A significant exception to the poor showing by the American manufacturers at this time was Glenn Curtiss, who was also one of the few early pioneers to retain a prominent position through the Great War, as World War I was known until some time after the start of World War II. Through his British licensee, White and Thompson, Curtiss came into contact with a former British naval officer, Lieutenant-Commander John Porte, who had earlier designed a glider and had also flown for the French Deperdussin Company. Porte joined Curtiss in the United States early in 1914. Together, Curtiss and Porte designed a large flying boat which they intended to be capable of transatlantic flight and which had been commissioned by Rod Wannamaker in an attempt to win a prize offered by the London *Daily Mail* for the first non-stop transatlantic flight. Normally referred to as the Curtiss–Porte *America*, the flying boat was a biplane with two inter-wing 90-hp Curtiss OX–5 engines, and the first of several aircraft was ready in June, 1914. The outbreak of World War I frustrated Wannamaker's ambitions for the flight, fortunately, since despite its many virtues and rugged reliability, the *America* was not a transatlantic aeroplane, even though the early requirement was to fly the

Newfoundland–Ireland leg non-stop. The idea of direct United States–Europe flight was some years away!

Porte returned to Britain after the outbreak of war in Europe in August, 1914, rejoining the Royal Naval Air Service as a squadron commander, a rank equivalent to squadron-leader in the Royal Air Force or major in the United States Army, and urged his superiors to test the *America*. They probably needed little persuading for, on 30th September, 1914, two Americas were shipped to Britain, designated the H.14 and made their first flights in November, 1914. Four more aircraft of this type followed, two of which were assembled by the Aircraft Manufacturing Company, while another two were built under licence by S. E. Saunders, at Cowes. These aircraft were powered by a variety of engines in an attempt to assess the suitability of different powerplants. They included Anzani, Clerget, Beardmore and Sunbeam products.

The operational debut of the *America*, or H.14, was not long in coming, with two of the first six aircraft being despatched to Gibraltar for maritime reconnaissance duties. The following year fifty H.14s were ordered, all using 100-hp Anzani engines, but the aircraft proved to be underpowered and, once more powerful replacements became available, they were relegated to the training rôle, although only three of these survived the war. Eventually, the H.14 was known as the Small America to distinguish it from another development, the Large America.

The development of the Curtiss-Porte line of flying boats divided at this stage. Curtiss developed and tested a development of the America in July, 1916, using two 160-hp Curtiss engines on the prototype, although the shortage of suitable American engines before production of the Liberty built up meant that production versions of the aircraft, known as the H–12, or Large America, had to use imported Rolls-Royce Eagle engines, a combination of airframe and aero-engine which resulted in a fine aircraft. Porte, on the other hand, was assigned to the Seaplane Experimental Establishment, which had been established at Felixstowe on Britain's east coast. There he did much to improve the original America design. The combination of H–12 wings, Rolls-Royce Eagle engines and Porte's own rather stronger hull resulted in the first of the Felixstowe flying boats, with an experimental F.1 followed by the production F.2, which was enough of an improvement on the H–12, of which the Royal Navy had received more than seventy, to be ordered into production with little delay.

Both the Curtiss and Felixstowe factories produced a series of fine flying boats. Curtiss built a smaller version of the H–12, the single-engined HS–1, with a 200-hp Curtiss VX3 engine, while a later version of this aircraft, the HS–1L (for Liberty) used a single 360-hp Liberty engine. The requirement for this machine came from the United States Navy, which needed a smaller flying boat which could be stowed more easily aboard a seaplane-carrier. Meanwhile, as they waited for their first HS series flying boats to be shipped to Europe, USN pilots in Europe flew Tellier and Donnet–Lévèque flying boats. The first HS flying boats arrived in Europe on 24th May, 1918. Later developments of the HS, the HS–2 and HS–2L, had an extra 12-foot wingspan, increasing the 360-lbs warload of the original design to 430 lbs, in addition to a Lewis gun in the front cockpit. More than 1,000 HS series flying boats of all kinds were built, with some 675 being built by Curtiss, with another 250 by LWF, eighty by Standard, sixty by Gallaudet, twenty-five by the then new Boeing Company and twenty-five by the USN itself at the Naval Aircraft Factory. Two HSs were built by another new American company, Loughead, which later changed its name to Lockheed. This was formed in 1916 by Allen and Malcolm Loughead, who had earlier, in 1913, designed an aircraft of their own with their friend, John K. Northrop. Curtiss was also building updated versions of his pre-war F-boat at this time, with 150 entering USN service as flying boat trainers.

Felixstowe, meanwhile, to give the 'brand name' of the Seaplane Experimental Station, continued working on its own series of aircraft, many of which were built in quantity by other manufacturers. As an example, with the F.2A, delivered between November, 1917, and May, 1918, 100 out of the 170 aircraft built came from the Isle of Wight factory of S. E. Saunders. The F.2 and F.2A were followed by the F.3 using 320-hp Sunbeam Cossack engines, though a number of production F.3s used the Rolls-Royce Eagle VIII engine. This flying boat was also slightly larger than the F.2, but only a hundred out of 260 aircraft ordered could be delivered before the end of hostilities, including a small number built by the Royal Dockyard in Malta. Generally, the F.3 was considered to be inferior to the F.2, and rather than being replaced by the F.3, many F.2s outlived the F.3s to be replaced by the F.5. The F.5 reversed the earlier trend for British engines to be sold to the United States, and in fact the F.5L was a Liberty-powered version, being built in the United States by the Naval Aircraft Factory, who were contracted to build some 500 examples of this aircraft, with another fifty being built by Canadian Aeroplane at Toronto and sixty by Curtiss, who were also building landplanes by this time, including the Jenny trainer. However, these orders were reduced at the Armistice, with the Naval Aircraft Factory order

Allen and Malcolm Loughead designed this aircraft with the help of John K. Northrop, before founding their own aircraft company, Lockheed.

reduced to 137 aircraft, that of Canadian Aeroplane to thirty and only Curtiss developing the type still further in the United States. It was an F.5L which dropped a torpedo of 400 lbs weight in trials on 22nd November, 1918.

There had also been some reorganization of partnerships in France before the outbreak of war. Lévèque had left Donnet–Lévèque, leaving management of the company in the hands of Denhout and Donnet, and in partnership with Louis Schreck he had formed Franco

One of two Lockheed-built HS–2L flying boats for the United States Navy in 1918.

British Aviation (FBA), although the use of 'British' in the title seems to have been little more than a marketing ploy with British contracts in mind, for which an office was opened in London. The first really successful FBA design was the Type B, a twin-seat biplane flying boat with a single 100-hp Gnome Monosoupape radial engine driving a pusher propeller, with a dozen of these aircraft being ordered by the Marine Nationale while the RNAS ordered forty-four from FBA, as well as another twenty built by Norman Thompson and a further sixty by the Gosport Aviation Company. Most Type Bs were relegated to training duties. The Type C, first flown in April, 1916, proved to be less successful and, using a 130-hp Clerget rotary engine, the aircraft was only ordered in small numbers by the French, Italian and Russian navies, mainly for training duties. A further development, the Type H, used either a 170-hp Hispano-Suiza engine or a 160-hp Lorraine-Dietrich, and while only small numbers of this aircraft entered service in the British, French, Belgian or Russian navies, almost a thousand aircraft of this type were built under licence in Italy by several manufacturers using a 160-hp Isotta-Fraschini engine. Many of the Italian aircraft were built after the Armistice, with the type remaining in production into the 1920s. Another development was the 200-hp Hispano-Suiza-powered Type S, mainly used by France and Belgium.

Apart from the Short Folder and the small Sopwith Tabloid, winner of the 1914 Schneider Trophy contest and available either as a landplane or as a floatplane, the other significant British seaplane of the war was the Fairey Campania. The Campania was named after the seaplane-carrier HMS *Campania*, and it was an aircraft of this type that first flew off this ship in June, 1916, using a trolley below its floats to take off from a launching platform over the bows, although the aircraft landed in the water in the conventional seaplane manner. Officially designated by Fairey as the F.16, the conventional single-engine and twin-float biplane Campania used either a 250-hp Rolls-Royce Eagle IV engine or a Sunbeam Maori II. Although obsolescent by the end of the war, a number of Campanias remained in RAF service for some years afterwards.

Shorts themselves added developments of the Folder, including the 184, the 225, 275 and 320, using Rolls-Royce Eagle or Sunbeam Cossack engines, largely depending on the availability of engines as industry struggled to cope with wartime demand. A number of Felixstowe flying boats were also built under licence by Shorts during and after the war.

On the opposing side, the Central Powers – Germany and the Austro-Hungarian Empire – also had substantial numbers of seaplanes

One of the first aircraft to be designed for shipboard use, the Short Folder biplane being lifted aboard HMS *Ark Royal*.

and flying boats at their disposal. One of the leading manufacturers was the Austro-Hungarian firm Jakob Lohner Werke, in Vienna, whose first flying boat was the biplane Type E, a twin-seat aircraft using two 85-hp Hiero engines, which enjoyed the distinction of making the first operational sorties by an Austro-Hungarian aircraft during August, 1914. Some 200 Type Es and a training version, the Type S, were built during the war. A development of the Type E was the Type L, a sesquiplane with a 140- or 180-hp Hiero or Austro-Daimler engine, two seats side-by-side in an open cockpit, and a warload of more than 400 lbs, in addition to a machine-gun in a rotatable mounting. Lohner built most of the 170 Type Ls ordered by the Austro-Hungarian Navy, mainly for operation against Italian shipping and land targets in the Adriatic. A reconnaissance version was the Type R, with three seats.

By accident, the Lohner Type L was to inspire a range of successful Italian flying boats, built by Nieuport–Macchi, the Franco-Italian

A later Short aircraft, the 310, launching a torpedo – while not commonplace, aerial torpedo attacks were a successful World War I innovation.

predecessor of Aeronautica Macchi. An Austro-Hungarian aircraft, L40, fell into Italian hands on 27th May, 1915, and after its capture was delivered to Nieuport–Macchi, with an order for a copy to be built for the Italian Navy. The first Macchi-built flying boat, the L40 copy was designated the L.1, to acknowledge its Lohner origins, and fitted with a 150-hp Isotta–Fraschini engine and forward-firing Revelli machine-gun. This aircraft was ready in June, 1915. So successful was the L.1 that a further 140 were built between June, 1915, and late 1916, although after this the type started to be replaced by FBA flying boats. A more lively development of the L.1 followed on the production lines, the L.2, with a lighter weight combined with a slightly more powerful 160-hp Isotta–Fraschini engine, although just ten aircraft were built to this design. The next design was more heavily influenced by Macchi's own design concepts; designated the L.3, this biplane featured an inter-wing mounted 160-hp Isotta–Fraschini V-4B engine which produced a 90 mph top speed, a range of 280 miles and a ceiling of 19,700 feet, with more than 200 L.3s built for the Italian Navy for use on reconnaissance, bombing and convoy protection duties. Before the Armistice, there were two more Macchi designs, the M.3 and M.5, both of which were single-seat fighter flying boats, with maximum speeds of up to 120 mph and two Revelli machine-guns. Some 240 M.5s were built for the Italian Navy, operating as escorts for bomber and reconnaissance flying boats. A still more advanced flying boat fighter was the Macchi M.7, although there were few of these in service at the Armistice and the aircraft saw most of its service during the post-war years, with the most eventful part of it being participation in the Schneider Trophy contests.

The Italian Navy also used a small number of Caproni and Savoia aircraft, including small Savoia flying boats, but the other significant Central Power's aircraft manufacturer was the German Friedrichshafen concern, with its FF series of seaplanes, which first entered service during spring, 1915, while successive marks of this aircraft were in service from 1915 to the end of the war. The biplane FF series were all conventional aircraft with a single engine driving a tractor propeller, and were the most extensively deployed German seaplane. Altogether, some 800 FF33, FF39, FF49 and FF59 seaplanes were used on coastal patrol, fleet observation and fleet co-operation duties, and anti-submarine warfare, operating from North Sea and English Channel seaplane stations, and from German warships. An FF33E, known as the *Wolfchen* or *Little Wolf*, was also operated from the German commerce raider *Wolf* in the Indian and Pacific Oceans, helping this German vessel to account for twenty-eight allied merchant vessels. The FF33 was the most successful Friedrichshafen machine, with some 500 built, while the FF39 which followed had a strengthened fuselage and a 200-hp Benz engine, as well as a rearward firing machine-gun. The next most successful aircraft from this concern was the FF49, of which some 240 were built, including a number by LVG, and this aircraft was the basis of a number of post-war civil conversions for up to five passengers in a cabin behind and below the pilot's open cockpit, operating in Germany and Denmark with a number of small airlines.

Although fighting alongside the Allies, there was little effective assistance from the Russians who were beset with severe economic and political problems during the war years, culminating in the October, 1917, Russian Revolution. A small number of Grigorovich M–5 and M–9 flying boats were available, but these did not become involved in any of the major battles of the war.

During the war, both sides established a chain of defensive and reconnaissance seaplane stations along their coastlines. As the Allies fell back into France, the German Naval Air Service, which had started with just thirty-six seaplanes at the outset of war, established a seaplane base at Zeebrugge, captured from the Allies in December, 1914. As the Germans advanced down the Belgian coastline, their most southerly seaplane station was ultimately established at Ostend. Eventually, the German seaplane stations stretched from Ostend on the Channel coast to the Friesian Islands, Heligoland and Sylt in the Baltic, while the British had their bases on the Isle of Sheppey and the Isle of Grain, as already mentioned, and in East Anglia, the location of Felixstowe, and further north along the North Sea coast. The Royal Navy also created a fleet of seaplane-carriers, including the converted Cunard liner HMS *Campania* and the former Channel packets *Ben-my-Chree* and HMS *Engadine*, with these ships becoming operational during 1915 and 1916. The French converted two captured German cargo vessels, the *Ann Rickmers* and *Rabenfell*, as seaplane tenders, adding two flying boat tenders, the *Nord* and the *Pas de Calais*, also converted Channel packets, before the end of 1916.

HMS *Campania* was amongst the first seaplane-carriers to be able to launch aircraft while under way at sea, using a 200-foot-long launching platform while aircraft took off with trolleys under the seaplane floats. Before long, large warships were being fitted with take-off ramps over gun turrets, followed later by catapults. Seaplanes, including the Campania, generally offered a maximum speed at 80–90 mph, while the Campania's range of three hours was also typical. The FF33 and FF39 were similar. Ships such as HMS *Campania*, which was built in 1893, suffered from many shortcomings, not the least of which was a 22-knot maximum speed, which meant that she was unable to keep up with the rest of the fleet and certainly could not hope to regain her position with the main battle fleet after stopping to pick up aircraft.

The seaplane and the flying boat – there were still very few amphibians in regular service – lacked the speed and the rate of climb to be able to offer an effective counter to enemy fighters, and the defensive armament of the flying boats was far from heavy or effective – that was still to come. Several attempts were made to overcome this problem, one of which entailed using a flying boat. A Felixstowe F.2 flying boat, or Baby America, or, as some would still call it, a Porte Baby, was test flown on 17th May, 1916, carrying a Bristol Bullet landplane scout, or fighter, biplane atop the upper wing of the F.2's mainplane, successfully launching the Bullet in mid-air. However, while this combined the speed, manoeuvrability and rate of climb of the fighter with the flying boat's endurance, the solution was too

Another early seaplane carrier, HMS *Engadine*, with a Sopwith 'Schneider', i.e. Tabloid, floatplane.

A Fairey Campania seaplane take-off from HMS *Campania*, of course! Note the launching trolley, simple, but effective!

impractical for regular use. A year later, in an attempt to extend the radius of action of the flying boat itself and provide more effective and long-range reconnaissance from warships on patrol, F.2 flying boats were towed into action aboard purpose-built barges behind destroyers, although there were problems with the stability of the barge and flying boat combination while under way at sea. The real solution to these problems lay in the aircraft-carrier with a variety of landplane types, although a reasonably successful and more immediate solution to the problem of fighter cover at sea lay in putting Sopwith Camel fighters aboard barges towed behind destroyers, and this was used on several occasions.

The seaplane even suffered problems in coping with the menace posed by the relatively large and slow-moving German Zeppelin airships, used for anti-shipping duties and bombing raids on British towns. Climbing slowly past a Zeppelin before making a strafing run, the pilot of a seaplane risked being shot down by defensive machine-gun fire from the airship, or of having the hydrogen-filled gas-bag blow up in his face if he was successful! Aircrew did not have parachutes at the time, not so much because there were no practical parachutes to be had, indeed, there were effective parachutes and further development could have been undertaken, but because of the callous attitude of many senior military commanders, who believed that the provision of parachutes would encourage aircrew to abandon their aircraft. Apart from the humanitarian principles involved, this also showed scant appreciation of the value of skilled and experienced aircrew or of the effect on morale. Indeed, the experience of World War II suggests strongly that parachutes did not encourage premature abandonment of an aircraft, and even today, with ejector-seats, one could argue that such safety measures actually encourage aircrew to expend further effort in saving an aircraft.

Co-operation between the Allies was not as strong in World War I as it was in World War II, but nevertheless, it did happen. For example, in December, 1914, two French Nieuport seaplanes flew reconnaissance missions in Egypt against Turkish forces, on behalf of the British forces in the area, operating from the seaplane tenders *Ann Rickmers* and *Rabenfell*. Almost a year later, six French FBA flying boats were dispatched to Venice to aid Italian forces in operations throughout the Adriatic, although later these aircraft were to prove easy prey for German Brandenburg fighter seaplanes.

The seaplane was not long in being used for combat. On 17th August, 1915, during the ill-fated Dardanelles campaign, a Short 184 seaplane from the seaplane-carrier, *Ben-my-Chree*, flown by Flight

Commander C. H. K. Edmunds of the RNAS, dropped a 14-inch torpedo to sink a 5,000-ton Turkish military transport in the Sea of Marmara, Turkey's inland sea between the Dardanelles and the Bosporus – the first successful aerial torpedo attack. A year or so later, the Germans were to return this somewhat dubious compliment when a German naval seaplane torpedoed a British vessel in the North Sea.

Bearing in mind the major rôle of the aeroplane during World War I, and the availability of the seaplane-carriers, the fleet deployment of seaplane-carriers and their aircraft occurred less often than one might have expected. This was doubtless due to the relative absence of fleet actions during the war, since, with the exception of the U-boat campaign, the one major naval battle was that at Jutland, which was both the first major naval engagement between modern capital ships and the first to involve naval aviation, although in a relatively small way. The British Grand Fleet, under Admiral Jellicoe in the battleship, HMS *Iron Duke*, with the seaplane-carrier, HMS *Engadine*, faced the German High Seas Fleet under Admiral Scheer in the battleship, *Friedrich der Grosse*, which did not possess the advantage of aerial reconnaissance. On 31st May, 1916, Flight Lieutenant F. J. Rutland, RNAS, and his observer, Assistant Paymaster Trewin, took off from the seaplane-carrier, spotted the German cruiser squadron and radioed its position back to *Engadine* at 15.10. On their return flight, a broken fuel pump forced Rutland and Trewin down into the sea, but after repairs they managed to take off and return to their ship. A battle cruiser action between the reconnaissance squadrons of the two fleets followed during the early evening, the inconclusive main battle taking place during the late evening.

Rutland's achievement was typical of the man. Shortly afterwards he received the rare honour of being awarded the Albert Medal in gold, for the rescue of a rating who had fallen between HMS *Engadine* and the crippled HMS *Warrior*, going down into the water between the two vessels at the not inconsiderable risk of being crushed. Unfortunately, the rating was found to be dead from exposure when recovered from the water.

Out of 140 U-boats sunk by the Allies during World War I, aircraft only accounted for five because the anti-submarine warfare (ASW) capabilities of the flying boat and seaplane were almost non-existent compared to the many specialized items of equipment carried by anti-submarine aircraft today. During World War I, an aircraft's main chance of destroying an enemy submarine was to find it on the surface or in clear shallow water. It might just be possible to see the silhouette of a U-boat below the surface, but the waters of the North Sea and the

English Channel are renowned for their murkiness. In the circumstances, it is perhaps not surprising that the first U-boat to be destroyed by an aircraft was not until May, 1917, when a Curtiss H–12 of the RNAS sank U–36 on 20th May. A few months later, on 28th September, U–32 was also sunk.

There were other successes, such as when the H–12 flying boat became the first US aircraft to destroy an enemy aircraft when, on 14th May, 1917, an H–12 shot down the Zeppelin L–22, the flying boat raking the airship with machine-gun fire until it fell into the North Sea. Another even rarer success for the flying boat occurred in June, 1918, when three Felixstowe F.2s and a Curtiss H–12 of the RAF flew into twelve Brandenburg seaplanes. Though the Curtiss was shot down and one of the F.2s had to land for repairs, this was at the cost of six of the Brandenburgs. However, this was flying boat against seaplane, against landplanes the story would have had a different ending.

Of course, not all of the development occurring to the flying boat and the seaplane as the war progressed was ready in time for active service, and there were many other minor types of flying boat and seaplane, including many converted hastily from landplanes, which contributed to the war effort. Aircraft such as the F.5 were to soldier on for some years, while the Russian Revolution was having an as yet unforeseen influence on the development of the flying boat and of air transport, forcing a talented young Russian designer to leave his homeland, eventually to settle in the United States.

The Porte, or Felixstowe, Baby America flying boat with the Bristol Bullet-fighter – an early attempt at improving the defensive capability of the flying boat.

3
THE SCHNEIDER TROPHY

The early years of aviation were dominated by the landplane. As we have seen, it was not until 1910 that the floatplane or hydro-aeroplane first appeared in powered form, with the tentative and courageous flight by the Frenchman, Henri Fabré. In 1911, the American, Glenn Curtiss, built and successfully tested the first practical floatplane, after which a limited number of floatplanes and flying boats began to appear, although for some years these remained less satisfactory in performance than their wheeled counterparts.

This relative neglect of the hydro-aeroplane disappointed the wealthy young Frenchman, Jacques Schneider, who was convinced that the key to the successful development of the aeroplane lay with the floatplane and the flying boat. The son of an armaments manufacturer, Schneider himself was a mining engineer by profession, although his interest in aeronautics had been first kindled by the demonstration flights of Wilbur Wright in France during 1908. Taught to fly by none other than Louis Blériot, in the same year that the great French pioneer had flown his landplane across the English Channel, Schneider also became interested in ballooning, and in 1913 he established a French altitude record of more than 30,000 feet in his balloon *Icare*. An old injury prevented him from taking a place in the forefront of the pioneering efforts of the early pilots.

Nevertheless, Schneider's contribution to the development of the hydro-aeroplane was to be immense. At a banquet held on 5th December, 1912, after that year's Gordon Bennett air race for landplanes, he presented a trophy, the Jacques Schneider Trophy for Hydro-Aeroplanes, to be contested for annually by national teams of hydro-aeroplanes. In common with the Gordon Bennett Cup, the Schneider Trophy contest was to be under the overall supervision of the Fédération Aéronautique Internationale, with competitors entered by national air clubs affiliated with the FAI. Qualifying tests were to be held in order to find a team of three entrants from each participating country, and the winning country of each year would assume the responsibility of hosting the following year's contest. Three victories by any one country in five years would qualify for an outright award of the Schneider Trophy.

Hardly an object of beauty, the Schneider Trophy consists of a base

Jacques Schneider, the Frenchman who instigated the Schneider Trophy contests after the 1912 Gordon Bennett race for landplanes.

of marble decorated with bronze, upon which a silver, nude, winged figure is depicted kissing a zephyr recumbent in the crest of a wave. This wave contains two other zephyrs and the head of the sea god, Neptune, all in silver. Many who see the trophy for the first time immediately assume that the winged figure must be Icarus, but this is to ignore the ample evidence of femininity! For those who have not seen it, there has sometimes been an inaccurate belief that the trophy is in fact a cup – the term 'Schneider Cup' crops up now and again – but this has never been so.

The first Schneider Trophy contest bore little resemblance to the major event of the post-World War I period; it was included, almost as an afterthought, in the Monaco Hydro-Aeroplane Meeting of 3rd to 17th April, 1913, organized by the International Sporting Club of Monaco. On the other hand, the trophy contest and its eliminating trial were the only really successful parts of the meeting. The trial was somewhat unnecessary because the number of would-be French contestants had been whittled down to three by the weather and inexperienced flying and rudimentary aircraft design left only one other competitor.

Most of the Monaco meeting was concerned with the preliminary trials for the main event, the hydro-aeroplane Grand Prix de Monaco, but out of sixteen aircraft only seven survived for the start. Miraculously, there was just one fatal accident. Only two of the three aircraft which eventually became airborne to start the race were intact at the end of it, and then only because they sought shelter. The race was finally abandoned in bad weather.

Not for the last time, the four aircraft competing in the first Schneider Trophy contest were conversions of landplanes, and all used floats designed and built by the floatplane pioneer, Henri Fabré. The Frenchman Maurice Prévost had a Monocoque Deperdussin, specially fitted with floats for the occasion. The almost equally famous French stuntman and long-distance flight pioneer, Roland Garros, had a Morane–Saulnier, while the third member of the French team, Dr Gabriel Espanet, and their American rival, and only non-French competitor, Charles Weyman, both used Nieuports. These aircraft were all monoplanes. Their task was to cover the ten-kilometre Grand Prix course running between Monte Carlo and Cap St Martin twenty-eight times, with the first half-lap spent taxiing on the surface to prove that the aircraft were true hydro-aeroplanes and not freak machines.

One of the early, albeit unsuccessful, Schneider Trophy contestants was the popular American aviator Charles Weyman (standing).

Another early Schneider contestant, Roland Garros, with friend.

This was to be no mean feat. Float design was primitive, although Weyman's Nieuport used a new design of stepped float based upon hydroplaning experience. Even if the take-off could be handled successfully, once in the air the floats and the tangled web of supporting struts and wires were not conducive to good aerodynamics. They decided upon an early start on the 17th April, at 08.00.

Prévost took off first, followed by Garros, who literally drowned his engine in spray while porpoising during the take-off run, and had to return to the slipway. Finally came Espanet and Weyman. Engine trouble in the eighth lap forced Espanet to retire, then Prévost, who had misread the rules, taxied over the finishing line instead of flying over. Procrastination while the stop-watch continued to run stole much of Prévost's time, for he refused to make a further final lap in the face of what seemed to be certain defeat by Weyman, only to find the American was forced to land in the twenty-fifth lap with an engine oil leak. After Weyman's trouble, Prévost relented, and flew a final lap around the course, but since his time for the race had run without stopping from his original take-off, the delay reduced his winning average speed from 61 mph to 45 mph. Garros, who had abandoned a second attempt while the contest was in full swing, returned to the start line for a third time and eventually finished in second place, although his speeds of up to 60 mph were of little use when averaged together with the time spent on the ground for repairs and indecision. Altogether, the event was sporting and leisurely, indeed, it must have been hard for many to take it seriously.

The first Schneider Trophy winner was Maurice Prévost, with a Monocoque Deperdussin floatplane monoplane.

French victory in 1913 meant that the Aéro Club de France was left to stage the second Schneider Trophy contest, and in fact they returned to the Monaco course of the first contest, holding the event on Saturday, 18th April, 1914. The 1914 contest was very much more in line with Jacques Schneider's original objective in sponsoring the trophy, with a truly international competition and a marked improvement in the performance of both men and aircraft. It also pointed towards the shape of the aeroplane over the next few years and foretold the ending of the French domination of high-performance aircraft development, which had endured from about 1910.

An 08.00 start was decided, and this time the competitors were blessed with good weather and a light breeze. The main departure from the rules of the 1913 contest was that the half-lap taxi run was replaced by two touchdowns, of no particular length, during the first lap. So much for seaworthiness!

The period immediately preceding the day of the contest had been hilarious for the French. These masters of the air had adopted a very superior attitude to the non-French competition which had been in no way diminished by their suffering a number of unwelcome surprises. Maurice Prévost's Monocoque Deperdussin failed to become airborne in a choppy sea during the eliminating tests, thus denying the French their most experienced pilot and most promising aircraft. The Deperdussin had had its 160-hp engine of the preceding year replaced by a 200-hp unit. Still, the French remained confident that their team, with Gabriel Espanet and Pierre Levasseur, both in 160-hp Nieuports, and Roland Garros in a 160-hp Morane–Saulnier, would be more than a match for the other entrants.

Only Charles Weyman, still with a Nieuport, was taken seriously by the French. The German Aviatik Arrow, flown by Ernest Stoeffler, was dismissed out of hand as being too slow, so was the Swiss-owned and flown (but French-built), FBA flying boat with a 100-hp Gnome

Victor of the 1914 Schneider Trophy, the Sopwith Tabloid, a small aeroplane with a high performance.

Monosoupape. Little attention was paid to one of the British entries, a Morane–Saulnier flown by Lord Carbery, and the other, a 100-hp Gnome Monosoupape-powered Sopwith Tabloid floatplane of remarkably small dimensions, was treated with amusement and ridicule by the French. Even worse ridicule was reserved for the other American contestant, William Thaw, who had entered a battered Curtiss flying boat. He was eventually shamed into borrowing a 160-hp Monocoque Deperdussin. In the event, the Aviatik was a non-starter, as would Carbery have been had he not been able to borrow a Monocoque Deperdussin.

The starting order was Levasseur, Espanet, Burri in the FBA, then Howard Pixton in the Sopwith Tabloid, who gave such a sparkling performance from the start that both Weyman and Garros decided that participation in the contest would be hopeless. A spell of engine trouble during the middle of the contest reduced the Tabloid's lead slightly, but the small machine soon recovered and impressed the hitherto cynical French with its speed and manoeuvrability. Carbery ran into difficulties during the alighting tests and wrecked the engine of his machine during a second start, while the Swiss, Burri, ran out of fuel on the twenty-third lap. An average speed of no less than 86.78 mph was recorded by the Tabloid during the contest, and as an encore the aircraft remained airborne to establish a hydro-aeroplane record of 86.7 mph for the 300 kilometres, then made a successful landing in a sea which had suddenly become rough as if to prove that this small aeroplane was no freak design but a thoroughly practical machine.

Naturally, the Schneider Trophy contest did not take place during World War I, and the right, or perhaps more accurately the obligation, to stage the first competition after the war remained that of the Royal Aero Club of the United Kingdom. The 1919 contest, held on 10th September, is on record as one of the worst organized, if not the worst, since the whole affair proved to be little more than a succession of errors, although fortunately no one was killed.

The first, and perhaps the biggest, mistake was to hold the contest at Bournemouth on the south coast of England, a resort which, despite having other attractions, completely lacked the facilities for a seaplane meeting. Competitors had to fly from their base at Cowes to Bournemouth, where the last minute preparations were put in hand on the morning of the race, amidst a throng of holidaymakers on the beach and in the water. A second mistake was to pay insufficient attention to the weather, and finally, there was again some confusion over the course and the starting procedure. Ten laps were to be made of a twenty-nautical-mile course, running from Bournemouth to Swanage, then to Christchurch and back to Bournemouth. Seaworthiness was to be proved by two alightings which had to be made during the first lap.

A complete team was fielded by Great Britain for the first time, with Vincent Nichol flying a Fairey IIIA, Harry Hawker a Sopwith seaplane and Basil Hobbs a Supermarine Sea Lion flying boat. The Fairey was a well-used aeroplane on which Dick Fairey had tested many different engines. (Later on the aeroplane was to be temporarily converted to an amphibian in 1920 for the Air Ministry contest to find a suitable aircraft for the RAF, coming third behind the Vickers and Supermarine contestants.) An Avro 539 seaplane was eliminated during the Schneider Trophy preliminary trials on 8th September. The French team consisted of a Nieuport and a Spad, and the first Italian Schneider contest entry was a Savoia S.13 flying boat, flown by Sergeant Guido Jannello.

The original intention had been to start the race at 14.30, but the mist of early morning had returned by 14.00, causing a postponement to 18.00. This was welcomed by the French, one of whose aircraft had sustained damage to its floats on landing at Bournemouth after the flight from Cowes. Further changes of plan then brought the starting-time forward to 16.30 and then pushed it back slightly to 16.45, and in the midst of this 'will we, won't we' confusion, the French team decided to withdraw from the event.

The three British entrants all took off within a short time of one another, in spite of some further confusion over the starting arrangements, and flew into dense fog at Swanage, as a result of which both the Fairey and the Sopwith abandoned the contest on returning to Bournemouth at the end of the first lap. The Sea Lion landed in Swanage Bay to find its bearings, sustained damage on taking off, and promptly sank on alighting at Bournemouth! Meanwhile, Jannello had taken off after starting correctly, and a decision over whether to abandon the contest was deferred while the Royal Aero Club officials waited to see whether or not he could complete the course. Jannello, however, appeared to be flying superbly, except for the final lap on which he appeared to make unusually good time. In order to avoid any risk of disqualification and for fear that he may have missed a marker-boat – he had in fact done just that at Christchurch – he was sent around for a further lap, but he ran out of fuel and had to make an emergency landing. It was at this time that the organizers also discovered that Jannello had been flying around the wrong course, using the reserve marker-boat moored in Studland Bay rather than the correct marker at Swanage. As a result he was disqualified, even though the mistake was an easy one to make and reflected badly on the organizers rather than on Jannello. Despite Italian protests, the race was declared void, and an attempt by the Royal Aero Club later to relent and award the trophy to Italy was overruled by the FAI, although in recognition of the misfortune which had befallen the Italian entry, Italy was allowed to stage the next contest.

The 1920 contest could only be an improvement compared to 1919, but even so, it was still far from being the type of success that Jacques Schneider had had in mind.

A shortage of funds meant that it was not possible for the British to send a team to Venice for the contest, while they also felt that they lacked a suitable aircraft. A French entry similarly failed to materialize. In part, this state of affairs owed much to a change in the rules which now required that all competitors carry 300 kilograms of ballast. This requirement, the result of Italian pressure, was dropped for 1921, and one could argue that the Italians had had the capabilities of their own aircraft in mind when formulating the requirement. As it happened, the Italians were lucky to be able to compete themselves, in spite of the absence of a competing foreign team. Weather and still inadequate design put two of Italy's three aircraft out of the competition, and not for the last time, one aircraft, a Savoia S.12 flying boat with Lieutenant Luigi Bologna of the Marina Militare Italiana, remained to fly ten laps of 37.2 kilometres each on 20th September. Determined not to falter during the contest, now a simple flight, Bologna naturally did not attempt to push the pace, and his average speed was only 107.22 mph.

A Savoia S.12 flying boat won the 1920 contest – it was the only aircraft to compete!

It was not surprising that many commentators of the time were becoming cynical about an annual non-event of such disappointing achievements. However, it was now the Italians' turn to attempt to win the trophy outright, and this desire reached almost fanatical proportions with no less than sixteen aircraft entering the eliminating tests for the 1921 Italian team!

The 1921 event was to consist of sixteen laps of 24.6 kilometres each, with alighting tests during the first lap, while Venice was to be the venue again, with the date fixed for 11th August.

The Italian team eventually consisted of two Macchi M.7 flying boats, to be flown by Giovanni de Briganti and Piero Corgnolino, and a Macchi M.19 flying boat, to be flown by Arturo Zanetti. France entered Sadi Lecointe in a Nieuport–Delage seaplane. During the contest, again without British participation, the float supports of the Nieuport–Delage collapsed during the alighting tests, while Zanetti's M.19 managed 141 mph on its first lap before catching fire and retiring, and Corgnolino's M.7 ran out of fuel on the last lap. Out of this debacle, Briganti's M.7 soldiered on at a fairly sedate 117.9 mph.

There had been speculation in Britain that, had the Gloster Bamel experimental aircraft, developed from the World War I Nighthawk fighter, been put on floats, it could have been entered for the contest and would have won. While this was a possibility, and no one can really be sure what would have happened given the many factors – the personalities, the weather and mechanical reliability – which had to be taken into account apart from the claimed performance of any aircraft, it seems unlikely, since the Bamel failed to win the landplane Coupe

This Macchi M.7 won the 1921 contest, flown by Giovanni de Briganti at an average speed of 117.9 mph.

Deutsch de la Meuthe, an event sponsored by the pioneer of the same name to replace the Gordon Bennett Cup after that trophy was won outright by Sadi Lecointe in 1920. One can only imagine that the Bamel would have been slower as a floatplane, burdened by the extra weight and the drag of floats, while the extra stress could have had some effect on engine reliability.

Fortunately, an event more worthy of the vision of Jacques Schneider was forthcoming at Naples in 1922, with thirteen laps of 28.5 kilometres each. The French and Italian teams were given government assistance for the first time in the history of the contest, but this did not prevent the French CAMS 36 flying boats from failing to meet the minimum requirements for entry, and left the Italians to face a purely private enterprise British rival. This sole British entrant was the Supermarine Sea Lion II flying boat, flown by Henry Biard. It had been redesigned and fitted with a new 450-hp Napier Lion engine by the young Reginald Mitchell, and so offered a more potent challenge than the Sea Lion of the 1919 competition. Biard faced an experienced Italian team, comprising Alessandro Passaleva in a Savoia S.51, Arturo Zanetti in a Macchi M.17 and Piero Corgnolino in a Macchi M.7. Like the Sea Lion, all of these aircraft were flying boats.

Unfortunately for the Italians, their best hope, the S.51, was damaged before the start of the contest after capsizing during the six-hour mooring-out test, which was a requirement of the 1922 contest, and while still able to compete, it did so at a distinct disadvantage. The event itself was timed to start at 16.00 on Saturday, 12th August; the heat of the Italian summer necessitating this late start.

Almost from the word go, the Sea Lion II showed that, for the first time in the post-war history of the contest, here was to be a victor worthy of the trophy, impressing the Italians not just by its speed but also by its exceptional manoeuvrability which enabled Biard to keep the aircraft close to the pylons at the ends of the course. All four aircraft completed the course, with the Sea Lion II's winning average being a close 145.7 mph against the S.51's 143.5 mph, suggesting that the S.51 might have won the contest had it not been for its misfortunes before the start: the following December, this aircraft established a flying boat speed record of 174.08 mph.

It was the 1923 contest at Cowes on the Isle of Wight that finally achieved the calibre of competition Jacques Schneider had had in mind, surpassing even the 1914 event. The Royal Aero Club had also learnt the many lessons of Bournemouth. The contest was fixed for 28th September. Navigability tests would be held on the preceding day, while the course itself would consist of five laps of 37.2 nautical

This Savoia S.51 came second in the 1922 contest at Naples, only 2 mph slower than the winning Supermarine Sea Lion II.

miles each. Originally, Britain, France, Italy and the United States – for the first time since 1914 – intended to field teams, but a last-minute withdrawal by the Italian team followed a loss of support from the Italian government. The British entry was reduced to a single Supermarine Sea Lion III flying boat – again to be flown by Henry Biard – by the failure of the other aircraft, the Blackburn Pellet, during the navigability tests. The Pellet was a World War I N.1B flying boat modified to incorporate a sesquiplane wing with a Napier Lion atop the upper wing. It could not be said to have been a completely new design, and its chances may have been limited anyway. The French entry had also been reduced to a single aircraft, the CAMS 38 flying boat of Lieutenant de Vaisseau Havel, after the CAMS 36 and Latécoère L.1 flying boats failed to start. Least depleted was the United States entry, but even their team suffered the loss in an accident of the Navy-built but Wright-designed NW–2 floatplane before the start of the event. This left two United States Navy Curtiss CR–3 seaplanes to be flown by Lieutenants David Rittenhouse and Rutledge Irvine, USN.

In excellent weather, Rittenhouse won at an average speed of 177.38 mph, while the other CR–3 came a close second at 173.46 mph, and the Sea Lion III trailed behind at 157.17 mph. At least the British entry managed to finish, unlike the CAMS 38, which was retired during the first lap while flying at the eastern end of the course near Selsey. It was obvious that the Europeans had nothing to compare with the CR–3 biplanes, with their 465-hp Curtiss D–12 engines, although the Sea

Lieutenant David Rittenhouse, USN, with his Curtiss CR–3 seaplane, which won the 1923 contest at Cowes.

Lion's 550-hp Napier Lion was obviously hampered by being placed in a flying boat instead of the faster seaplane. Clearly, in common with the French in 1914, the British in particular had underestimated the weight of the challenge being mounted by their opponents, even to some fairly high and mighty criticism of the length of the floats fitted to the American aircraft – almost as long as the fuselage and clearly unsatisfactory, was the view before the contest. Yet, the Americans had re-emerged as the leaders in aviation again, and, by comparison, the Europeans had put obsolescent aircraft into the contest.

A variety of circumstances combined to ensure that none of the European nations sent teams to Baltimore in an attempt to regain the Trophy in 1924, and the Americans could legitimately have flown over the course themselves, as the Italians had done in 1920, and won the Trophy for a second time. Instead, the National Aeronautical Association, which was the responsible body for aviation competitions in the United States, made the offer to the Royal Aero Club, which had been the last European body to withdraw a national team from the event, that the 1924 contest be cancelled. Needless to say, this was gratefully accepted.

Britain, France and Italy had all planned to send teams to Baltimore, even though the Americans had, if anything, increased their technological lead since their victory. In Italy, neither Cantieri Rivniti dell'Adriatico (CRDA) nor Piaggio were having success with their aircraft, while neither Supermarine nor Gloster in Britain were able to produce aircraft – the Supermarine Sea Urchin being abandoned because of difficulties with the drive from the hull-mounted engine to the propeller, while the more promising Gloster II suffered an accident, nearly killing its pilot, Hubert Broad. The USN itself had decided to concentrate on the Schneider, at the cost of leaving participation in the Pulitzer Trophy, a landplane contest, to the United States Army Air Corps, making the generous gesture by the Americans so much more of a sacrifice.

Baltimore was again the venue for the 1925 contest, also organized by the National Aeronautical Association. The course, in Chesapeake Bay, was to be completed in seven laps of 50 kilometres each. The preliminary trials were to be followed by a six-hour mooring-out period before the event, in an attempt to eliminate unseaworthy, and therefore freak, machines. Originally scheduled for 24th October, the actual event was postponed until the 27th because of the adverse weather which had persisted throughout the preparations for the contest.

Lieutenant James 'Jimmy' Doolittle, USAAC, winner of the 1925 Schneider contest with his Curtiss R3C–2 seaplane.

The American team consisted of Lieutenant James Doolittle of the United States Army, Lieutenants George Cuddihy and Ralph Ofstie, United States Navy, all of whom flew Curtiss R3C–2 seaplanes with 600-hp Curtiss V–1400 engines. Italy had entered two Macchi M.33 flying boats, to be flown by Giovanni de Briganti and Riccardo Morselli. Great Britain had entered Henry Biard with a Supermarine S.4 cantilever monoplane seaplane, and Bert Hinkler and Hubert Broad with Gloster III biplane seaplanes. Engine trouble caused the withdrawal of Morselli's Macchi, the S.4 crashed during trials because of wing flutter, and Hinkler's Gloster III was damaged by severe wave buffeting while landing for the navigability trials. The British had a Gloster Bamel on floats as a reserve aircraft. All of the British aircraft were powered by 700-hp Napier Lion engines, while an indication of the effects of such a growth in engine horsepower was the offset position of the starboard float on the S.4, to counter the effect of engine torque and the thrust of the propeller of the single-engined aircraft. This was increasingly apparent in the tendency for one of the floats of single-engined seaplanes to 'dig in' under full take-off power, leading to serious difficulties in building up take-off speed and retaining control. Also, of course, it was potentially very dangerous as take-off speeds increased. The British had also now abandoned the flying boat in a high-speed form, preferring the seaplane, although the Italians continued to cling to the flying boat for high-speed operation. Other problems were found with the new cantilever wing form of aircraft such as the S.4 and M.33 and gave rise to concern about wing flutter, a reflection of the still limited structural knowledge of the day, while the pilot's view from the S.4 was supposed to be 'appalling'.

The end result was simply a postponement of what would have happened a year earlier. Although Ofstie was forced to retire during the sixth lap and Cuddihy followed suit in the final lap, Doolittle raced home with an average speed of 232.57 mph, leaving Broad's Gloster trailing behind with a slow 199.17 mph and Briganti's Macchi still further back with a hopeless 168.44 mph.

Still denied Government support, even though test aircraft bought by the Air Ministry were to be made available to the manufacturers if they cared to sponsor an entry in the Schneider Trophy contest, the British team stayed at home in 1926 to lick its wounds. British hopes that the contest would be postponed once again were dashed by an Italian entry, although eventually the FAI ruled that after 1927, the contest would become biennial. Defeat in 1925 had been total. As an encore, on the day after his victory, Doolittle had established a new world speed record for seaplanes of 245.71 mph. The landplane version of the R3C–2, the R2C–1, had already set up what the Americans had regarded as a 'disappointing' 248.97 mph before the Schneider contest, especially since they had believed, quite realistically, that replacing wheels with floats would have cost the aircraft some 30 mph in top speed.

However, in one sense, defeat had been good for the British, stimulating fresh thought and renewed effort. They would in fact have been hard-pressed to field a worthwhile team in 1926, with three basic designs almost, but not quite, ready. It was on the first of these, the Gloster III development, that Hubert Broad, granted leave from his contract to de Havilland, was able to test-fly during August and September, 1926, with the Gloster IIIA and IIIB testing different wings and radiators in preparation for a new Gloster IV biplane. The other two designs were the Napier Lion-powered Supermarine S.5 monoplane and the 800-hp Bristol Mercury-powered Short Crusader monoplane.

Faced with staging the contest in 1926, the Americans chose Hampton Roads, Virginia, and a 50-kilometre seven-lap course into Chesapeake Bay and past Newport News to the naval base at Norfolk. Inspired or pressurized by dictator Benito Mussolini, Italy was able to provide the sole challenge to the United States. However, the fact that the Italians had full Government support cannot detract from the fact that the designer, Mario Castoldi, designed and built the Macchi M.39 monoplane seaplane in just six months. It did not help the Americans that they were having some bad luck, with a series of accidents losing them some of their best pilots, and problems also arose in the handling of the aircraft into which they had fitted a variety of powerplants.

Three M.39s with 800-hp Fiat A.S.2 engines were used by the Italian team sent to Hampton Roads. Their pilots were Major Mario de Bernardi, Captain Arturo Ferrarin and Lieutenant Adriano Bacula. The United States team, drawn entirely from the United States Navy on this occasion, used three Curtiss biplanes: the R3C–2 with a 600-hp Curtiss V–1400 engine, flown by Lieutenant Frank C. Schilt; a Curtiss R3C–3, flown by Lieutenant Walter G. Tomlinson and using a 700-hp Packard 1A–V–1500 engine, and a 700-hp Curtiss V–1550–powered R3C–4, flown by Lieutenant George T. Cuddihy. A heavy landing during the navigability tests caused the loss of Tomlinson's R3C–3, and the only available substitute was a standard Curtiss Hawk floatplane. The strength of the Hawk was its reliability: it could be counted on to finish the contest, which was more than the temperamental racing machines could promise. Also with reliability in mind, Castoldi's monoplane used a braced rather than a cantilever

Major Mario de Bernardi, winner of the 1926 contest with his Macchi M.39; the return of the monoplane to the list of trophy winners.

wing, to avoid the problems which had beset the S.4 and the M.33 the preceding year.

Although the 1926 contest was held on 13th November, later in the year than was usual for the Schneider contests, weather on the day of the contest was perfect. The first aircraft took off just after the starting time of 14.30. Both teams suffered aircraft damage: Cuddihy in the Curtiss R3C–4 retired before the end of the final lap with a fuel pump failure, and long before this Ferrarin had been forced to retire during his fourth lap with an oil pipe fracture. Bernardi's winning average speed of 246.50 mph was comfortably ahead of Schilt's 231.36 mph, while Bacula was lagging behind the leaders with 218 mph and Tomlinson's Hawk displayed the performance difference between service and racing machines with a lingering 137 mph!

Italy had regained the Schneider Trophy, and the monoplane had returned to the predominant position in high-speed flight. However, Italian determination to regain the trophy had to face a British challenge, backed for the first time by the British Government with Royal Air Force personnel flying and maintaining the machines. Meanwhile, the end of Government support for the Americans had been a not entirely unexpected blow to that country's hopes of regaining the Schneider Trophy in 1927. The last straw was the failure of the private-venture Kirkham-Packard design to produce a worthwhile performance during tests in America. In the face of Italian

Less successful was this Macchi MC.67 entered by Italy for the 1929 Schneider contest.

and British opposition to a month's postponement of the contest – unfairly the Americans might have been justified in thinking, after their earlier generosity – the United States had to withdraw.

So, the 1927 contest remained a bilateral rather than a truly international event. As at Hampton Roads the course consisted of seven laps of a 50-kilometre circuit. Venice was the venue once again.

The Italians were banking on a development of the M.39, the M.52, with a 1,000-hp Fiat A.S.3 engine, to defend the Trophy successfully. Three of these aircraft were ready, to be flown by Major Mario de Bernardi and Captain Arturo Ferrarin from the winning team of 1926, and Captain Frederico Guazzetti. Guazzetti's machine's A.S.3 was replaced by an A.S.2, giving the Italian team one truly reliable machine, a move prompted by problems during the testing of the A.S.3.

Great Britain's challenge was mounted by Flight Lieutenants S. N. Webster and O. E. Worsley, RAF, flying, respectively, propeller-geared and ungeared versions of the Supermarine S.5 seaplane, a braced cantilever development of the ill-fated S.4, and both using 875-hp Napier Lion engines. A third British aircraft was the biplane Gloster IVB, to be flown by Flight Lieutenant S. M. Kinkead, RAF. The Gloster also used a Napier Lion engine, driving the propeller through a reduction gear. A third British type, the Short Crusader monoplane, with an air-cooled radial engine, was taken to Italy for further evaluation but eliminated itself well before the contest by crashing as a result of having been reassembled after shipment with the aileron control cables reversed. Fortunately, the pilot escaped serious injury. The reversal of the aileron controls was easy to see when the aircraft was salvaged after a week from the water into which it had crashed – the magnesium alloy engine crankcase had dissolved leaving the aileron cables to view. The loss of the Crusader itself was far from being a tragedy, for not only had the pilot escaped, but the aircraft was already considered the slowest of the types assembled by Britain for the contest.

After a twenty-four hour delay due to bad weather, the contest finally started at 14.30 on Monday, 26th September. The Italian team was fated to lose from the start, as the unreliability of the up-rated engines forced Bernardi to retire during his second lap and Ferrarin during his first. Even Guazzetti, with his 'reliable' aircraft, was forced down after five laps with a broken fuel line. Only Kinkead, who had been doing sufficiently well at one stage to have challenged the leading S.5, had to retire from the British team with a cracked propeller spinner.

This sad tale of retirements and dashed hopes nevertheless meant a

The Supermarine S.5 regained the Trophy for the United Kingdom in 1927 – Webster the pilot stands with his hand on the aircraft, and designer Reginald Mitchell is in the centre of the front row.

happy ending for Webster, who won the trophy for Great Britain with an average speed of 281.65 mph. Worsley followed with a speed of 273.07 mph. Webster had lost count of the number of laps flown and discovered on landing that he had flown an extra lap! The high speed of Webster's aircraft was not enough of an advance to establish a new official record, even though it did beat the then existing speed record by 3 mph.

Although based on Calshot, at the mouth of Southampton Water, the 1929 contest actually had its start and finish lines 'drawn' at Ryde on the Isle of Wight. As in earlier years, seven 50-kilometre laps were required from the competitors, and taxiing tests were added to the mooring-out trial.

Again, this was to be another two-nation event, although by accident rather than by design. The Italians, taking a leaf out of Britain's book, attempted to obtain a number of competing designs from which a Schneider challenger or challengers could be selected. However, they squandered scant resources amongst Macchi, Fiat, Savoia–Marchetti and Piaggio and, in the end could still only race the Macchi design. The French were forced to withdraw because of a number of serious accidents during training for the contest in France, and in the United States poor Alford Williams was still plagued by technical problems. Not that everything went as expected for the British, whose Gloster VI monoplane had to be withdrawn because of problems with the fuel feed system to its Napier engine.

In spite of such widespread trials and tribulations, both Britain and Italy were able to field complete teams. Britain once again had Government support and RAF personnel for the contest, with Flying Officers H. R. Waghorn and R. L. R. Atcherley, RAF, both using 1,900-hp Rolls-Royce R-powered Supermarine S.6s, developed from the S.5. One S.5 was entered to be flown by Flight-Lieutenant D. D'Arcy Grieg, RAF. The Italian team consisted of two Macchi MC.67s with 1,800-hp Isotta–Fraschini engines, to be flown by Lieutenants Remo Cadringher and Giovanni Monti, and a modified Macchi M.52R with a 1,000-hp A.S.3 engine, to be flown by Warrant-Officer Dal Molin. A popular member of the 1927 team, the South African Flight-Lieutenant S. M. Kinkead, had been killed while flying in an S.5 on 12th March, 1928, in an attempt on the world's air speed record. The Chairman of the House of Commons Aviation Committee, Sir Samuel Hoare, paid tribute to Kinkead as 'a young man with an unrivalled record who might in the ordinary course have reached the highest post in his great profession'. The Inspector of Accidents concluded that Kinkead may have crashed either as the result of a stall or, in the sunlit

Testing! The eventual victor was the Supermarine S.6B, seen here on pre-contest checks at Calshot.

haze of afternoon in the Solent, had flown into the water. On the other hand, the Italians had the benefit, not to say the boost to morale, of having had Major de Bernardi fly his Macchi seaplane at Venice in late March to a new record of 320 mph, at a time when most fighters could still not manage 200 mph.

On 7th September, the day of the contest, the excellent weather of the preceding few weeks continued, but so did the bad luck of the Italian team. Even with their engines running at reduced power, both the MC.67s had to retire during their second laps, their pilots temporarily overcome by smoke and fumes. Another unfortunate was Atcherley, who lost his goggles on take-off and, with his head consequently well down inside the cockpit, took a pylon on the inside and was disqualified, although his speed records for 50 kilometres and 100 kilometres were allowed to stand.

Waghorn won the contest convincingly with an average speed of 328.63 mph, against Dal Molin's 284.4 mph and Grieg's 282.11 mph. Before his disqualification, Atcherley had achieved an average speed of 325.54 mph. The result was a pleasant surprise for Waghorn, who ran out of fuel on his last lap and landed thinking that the Italian team had won, only to find that he had repeated Webster's mistake of 1927 and flown an extra lap! However, Italy would have regained the trophy had

Waghorn failed to win, since the S.5 was not an adequate match for the Macchi M.52R. The Italians had been plagued with trouble before the contest, losing a Savoia–Marchetti S.65 tandem 'pusher and tractor' propeller twin-boom seaplane, with its pilot Captain de Motta, during a training flight. The British, making the most of things while their luck still held, had the RAF's High Speed Flight make an attempt on the world air-speed record. The Gloster IV was given the first try, and with Flight-Lieutenant Stainforth at the controls reached 336.34 mph, but Orlebar in the S.6 managed to push the record up to 355.8 mph.

The 1931 Schneider Trophy contest was an anticlimax. Plagued by troubles on their new Macchi–Castoldi MC.72, the Italians could not compete. This left only the British team, and on Sunday, 13th September, 1931, again at Calshot and over the 1929 course, Flight-Lieutenant J. N. Boothman, RAF, flew a Supermarine S.6B with a 2,350-hp Rolls-Royce R engine to win the trophy outright for Britain with an average speed of 340.08 mph. Later the same day, Flight-Lieutenant G. H. Stainforth took the S.6B to a world air-speed record of 379 mph.

The real drama for the 1931 Schneider Trophy was political and financial. Despite a promise in 1929 that the Government would support another British attempt on the trophy, the new Labour Government refused to provide financial assistance. A financial crisis was blamed for this short-sighted attitude, although the Labour Party has never been notable for understanding the needs of, and the value of, the aircraft industry. Promises of funds by the industry and by other businesses fell far short of the amount required, and it was not until Lady Houston, the widow of a wealthy shipping magnate, offered £100,000 towards the cost of Britain's entry, that the British team was able to make a start on its preparations for the event. It was another struggle to obtain the services of RAF personnel for a third event, but this too succeeded in the end. Costs had risen, and the trophy contest was no longer a sport for individuals or a marketing exercise for companies but a question of national prestige: the £2,000 of a Gloster II biplane, for example, had risen by 1929 to £25,000 for a Gloster VI, and two years later, costs were even higher.

In one respect, the Italians had the final word. On 23rd October, 1934, Warrant Officer Francesco Agello flew the Macchi–Castoldi MC.72, with its contra-rotating propellers, to an as yet unbeaten speed record for propeller seaplanes of 440.68 mph. The contra-rotating propellers, by balancing the thrust, were the ideal answer to the problem of one of the floats 'digging in' during take-off. Yet, from the S.6B, its designer, Reginald Mitchell, was able to develop the

The ultimate speed record for propeller-driven seaplanes was gained by this Macchi–Castoldi MC.72 in October 1934

landplane Supermarine Spitfire fighter, which formed the backbone of Britain's fighter defences during the Battle of Britain and served throughout World War II on every front on which British forces were deployed; thus the United Kingdom gained something lasting from the contests, apart from the trophy. It is a sobering thought that the Spitfire owns its direct and illustrious ancestry to a generous and patriotic individual gesture, without which Mitchell would have been denied his own contribution.

Jacques Schneider, born in January, 1879, died in 1928 in some poverty. Nevertheless, although he failed to see the final glory of the contest inspired by his trophy, on more than one occasion during his life, the trophy was won by aircraft which had a profound bearing on aeronautical development. The trophy itself remains in the care of the Science Museum, London.

4
SPREADING WINGS

The end of World War I brought peace to Europe and effectively ended Germany's role as a major aircraft manufacturer for many years, forcing German designers to work outside of their own country. For the Allies, obviously no such constraints applied, but there were to be other economic, and even political, problems for the still young naval air arms in an uncertain world. Youngest of the various air arms was, of course, the new Royal Air Force, which had taken over all British service aviation from the Royal Naval Air Service and the Royal Flying Corps, and which itself underwent a dramatic and sudden slimming down and reconstruction after the end of the war, largely with the objective of freeing the new organization of the influence, rivalries and prejudices of the older air arms which it had replaced. On the other side of the Atlantic, in the United States, a bitter battle to follow the British example and establish an autonomous air arm was to result in fierce exchanges between senior officers in the United States Navy and the United States Army Air Corps, showing just how wise the British concept of reduction and then reconstruction had been. The American controversy eventually got completely out of hand, with the court martial of the senior protagonist for an autonomous air service, Brigadier General William 'Billy' Mitchell, for insulting the United States Navy!

Friendly inter-service rivalry and *esprit de corps* can be an excellent thing, but bitter feuding does nothing for those involved and even less for those who are depending on efficient armed forces for national security!

Immediate post-World War I development of the maritime aeroplane by the navies and air forces of the world largely centred around existing wartime designs – many of which had proved themselves under arduous combat conditions. In the United States, where maritime air power was more firmly wedded to the concept of shore-based flying boats for long-range maritime reconnaissance than in Britain – where the idea of the seaplane and aircraft-carrier held sway – the developments of wartime aircraft continued to enter service. The Naval Aircraft Factory at Philadelphia was building the most up to date of the Curtiss line, the twin-engined H–16 with 345-hp Rolls-Royce Eagle engines or, as an alternative, the American-designed Liberty engine, as well as the smaller Curtiss HS–2 which tended to be used more for training and coastal reconnaissance duties – these aircraft had a price tag of $37,850 and $16,900 respectively. The Naval Aircraft Factory was also building the British-designed Felixstowe F–5L, using Liberty engines (hence the 'L' in the designation). By this time, the power of the flying boat meant that there was sufficient horsepower to spare for adequate armament protection against fighter aircraft, and the H–16 mounted no less than five Lewis guns, usually with one in the bow position, one or sometimes two in the rear cockpit, and one each in two waist openings on either side of the fuselage amidships. The F–5L carried a similar armament, as well as up to four 230-lb bombs or depth-charges. A contemporary of these three aircraft was the MF, or Modified F, also from the Curtiss design team and a development of the original Model F small flying boat of 1914. In its F and MF forms it had served the USN throughout the war years as a flying boat trainer. Curtiss built twenty-two MFs post-war and the Naval Aircraft Factory built another eighty.

The H–16 and the F–5 series in their various forms were to be the mainstay of the RAF and the USN for several years after the end of hostilities, and indeed the layout of these aircraft was to be the 'classic', or conventional, flying boat shape for the first generation of post-war aircraft. Doubtless the acceptability of the design was aided by the continuation of production, albeit in very small numbers, and the retention of many aircraft in service for several years. Not to put too fine a point on it, after the boom of the war period, many of the aircraft manufacturers were soon to find themselves in considerable difficulties, and aircraft production slumped from being one of the largest and most important, as well as one of the most modern, industries to being one of decline and insecurity. Several famous manufacturers nearly went out of business, either voluntarily as they sought other more profitable opportunities or involuntarily as bankruptcy loomed. Their workforces reduced in numbers although, under pressure as many were, the much-prized design teams were kept intact for the most part, not least because it can take as much design work for a low-volume production aircraft as it does for one which enjoys large orders.

This is not to suggest that all was lost for the pioneering companies. Short Brothers continued to produce the F.5, and in their hands it survived the demise of the Seaplane Experimental Station, which had founded the 'Felixstowe' line. Shorts produced, amongst other things, a metal-hulled version of the F.5 in 1925, which was the first metal-hulled flying boat in the world, in an attempt to overcome one of the persistent problems of the flying boat hull, the lack of adequate watertightness. Built more lightly than the old 'wooden wall' sailing ships, the flying boat hulls had to withstand a tremendous buffeting at speed during take-off and landing, and frequently sprang a leak or became waterlogged.

Not all was doom and gloom worldwide. Contraction of service aviation in Europe was not matched in the Far East; in Japan the Imperial Japanese Navy was rapidly building up its aeronautical expertise and its strength in the air, buying nine Felixstowe F.5s from Shorts. These aircraft joined a force of Schreck FBA.17 and Tellier flying boats, and were soon followed by a number of aircraft from other British manufacturers, including Vickers and Supermarine, as well as by many landplanes. Shorts also built an F.3 at this time for a private owner, an Australian millionaire, who never used it, and after the aircraft was delivered, rather than being used as some form of flying yacht it suffered the indignity of being used as a fisherman's beach shelter in Sydney.

In the United States, the new Bureau of Aeronautics, or BuAer, established by the United States Navy in 1922, introduced new designs, and cven before this introduced new aircraft designations, so that the F–5L, for example, became the PN–5, or Patrol Navy 5. The PN–5 was soon joined by a development, the PN–6, which had first flown as the experimental F–6L also originating from the Naval Aircraft Factory, and this type was to remain in service until 1928. A series of experimental machines followed throughout the 1920s. The first of these was the PN–7, using two 525-hp Wright engines and a new wing design, while the next aircraft, the PN–8, used two 475-hp Packard engines and was of all-metal construction. A variation on this design appeared as the PN–9, with modifications to the tailplane. Other aircraft included the PN–10 and the wide-hulled PN–11, while the PN–12, which incorporated many PN–10 features, marked the end of the purely experimental aircraft in this series, entering production at the Douglas, Keystone and Martin companies as the PD–1, PK–1 and PM–1, or Patrol Douglas, Patrol Keystone and Patrol Martin. Douglas built twenty-five PM–1s, Keystone eighteen PK–1s, and Martin thirty PM–1s, followed by twenty-five PM–2s, a twin-fin development of the original design. The aircraft in this series also showed some considerable improvement in performance. One of the earlier designs, the PN–5, could fly at 90 mph and had a ceiling of 5,500 feet, with a range of 830 miles, using two 400-hp Liberty 12A engines in an aircraft with a wingspan of 103 feet and a fuselage length of 50 feet. The PN–10, which first flew in November, 1926, had metal wings and twin 525-hp Packard engines, giving an endurance of 950 miles in just over eleven hours, although in 1927 this aircraft flew 1,569 miles in just over twenty hours with a load of 1,100 lbs. The PN–12 used twin 525-hp Wright Cyclone engines or Pratt and Whitney engines, of the same horsepower rating, and the prototype of this aircraft flew in December, 1927.

The United States Navy flying boats made many long-range flights, partly in the interests of setting records and earning prestige but also because of one of the real benefits of the flying boat during the 1920s, the ability to offer a longer over-water range than the landplane. The flying boat also allowed designers to overcome many of the problems which inhibited the development of larger landplanes, mainly because designers could create larger aircraft than undercarriage design could cope with. In addition, the flying boat offered a better chance of survival than the landplane in the event of an engine failure during an over-water stage. Designs by the Naval Aircraft Factory, Curtiss and Boeing all participated in attempts to span the longer over-water distances of the Caribbean and then, later, the Pacific.

However, immediately the flying boat had become a viable proposition, one of the earliest objectives for the pioneers became a crossing of the North Atlantic. After all, over the long ocean distances the aeroplane could show itself to far greater advantage compared to surface transport than on overland flights, where the aeroplane had to compete, in Europe and North America at least, with fast express trains which were sometimes almost as fast as the aeroplane and always more reliable and more comfortable. A British newspaper, the *Daily Mail*, fuelled the enthusiasm of the pioneers by offering a prize of £10,000 for the first non-stop North Atlantic flight, initially announcing this award in 1913. It was this prize which attracted the attention of the millionaire aviator Rod Wannamaker to the challenge of the North Atlantic, and encouraged him to commission the Curtiss–Porte flying boat for a transatlantic flight. However, as we have seen, before the aircraft was ready World War I intervened.

Wannamaker lost interest in transatlantic flight during the war years, but Glenn Curtiss did not. However, the four Curtiss NC (Navy Curtiss) flying boats were not intended to make the non-stop crossing specified by the *Daily Mail*, and so it was to a landplane, a Vickers Vimy bomber, flown by two Britons, Captain John Alcock and Lieutenant Arthur Whitten Brown, which flew non-stop from Newfoundland to Ireland in 1919, that the prize was awarded. Officially intended to be an anti-submarine flying boat, the first

The USN Curtiss NC–4 flying boat at Plymouth after its epic transatlantic flight in 1919.

aircraft in the NC series was the aptly designated NC–1, originally intended to be powered by three engines, but in fact first flown on 4th October, 1918, with four 400-hp Liberty engines, although the aircraft also offered a significant advance over the F–5 and H–16 types, with a range of 1,470 miles, a span of 126 feet and a fuselage length of 70 feet. The four engines were mounted in two inter-wing nacelles, each nacelle having a tractor and a pusher propeller, while the fuselage was itself a large nacelle, with the tailplane pointed on outriggers. On 25th November, 1918, the NC–1 set a world record by taking fifty-one people into the air. The next three aircraft, NC–2, NC–3 and NC–4 were also fitted with four engines, but in three nacelles in the inter-wing position with the middle nacelle containing two engines. The NC–2 first flew on 12th April, 1919, the NC–3 followed on 23rd April, and the NC–4 on 30th. The original aircraft, NC–1, did not take part in the transatlantic attempt, leaving the three later aircraft to take off from Trespassing Bay, Newfoundland, on 16th May, 1919, to fly the 1,400 miles to Horta in the Azores, with the formation under the command of Captain John Towers, USN, who flew in the NC–3. Unfortunately, the NC–3 force-landed 200 miles from Horta and had to taxi to this first landfall. NC–2 was even less fortunate, with the aircraft crashing in heavy seas some 100 miles west of Flores, although the crew were saved by a weather ship, one of a chain of USN and United States Coast Guard vessels stretched across the Atlantic to save the flying boat crews in the not so unlikely event of a forced landing.

But the NC–4, flown by Lieutenant-Commander Albert Reed, reached Horta and went on for Ponta Delgada, eventually reaching Lisbon on 27th May, completing the first transatlantic crossing. However, Reed then took his aircraft on to Plymouth, arriving there on 31st May, to be escorted into harbour by three Felixstowe F.2A flying boats of the Royal Air Force. Although the flight was certainly not non-stop, the achievement of the NC–4 was noteworthy nonetheless.

On its return to the United States, the NC–4 toured the eastern and southern United States and flew up the Mississippi to St Louis before finally being retired to the Smithsonian Institution, in Washington, where it was put on display in the National Aeronautics and Space Museum. A further six NC series flying boats were built, but not by Curtiss, since these aircraft were constructed at the Naval Aircraft Factory. They initially had three engines each, although they were later modified by the addition of a fourth engine. However, their service life was very short, being retired in 1922.

The PN–9 attempted another epic flight, the 2,200 miles non-stop from California to Hawaii – now one of the states of the Union but even at that time an important naval base. The arrival of the PN–9 in Hawaii was even stranger than that of the NC–3 in the Azores. Taking off from San Francisco on 31st August, 1925, the aircraft flew westwards for 28½ hours, covering 1,840 miles of its journey against strong headwinds and establishing a new non-stop distance record before running out of fuel and landing in the sea some 360 miles short of its destination, the giant United States Navy base at Pearl Harbor. Fortunately, the PN–9's hull remained watertight, and with the crew using the wing fabric as sails, the aircraft eventually reached Hawaii after ten days sailing!

An alternative to the PN–9 for the Hawaii flight was the PB–1, or Patrol Boeing 1, known to the manufacturer as the Model 50, which was completed in August, 1925. This aircraft had a metal hull and a wooden deck, with the fabric-covered wings usual at the time. It was designed to operate as a long-range maritime reconnaissance flying boat with three Lewis guns and 4,000 lbs of bombs, while powered by two 800-hp water-cooled Packard engines in a single inter-wing nacelle with tractor and pusher propellers. However, the aircraft suffered engine difficulties before the Hawaii flight could be attempted, and so the opportunity of making this fell to the sturdy PN–9. As a result, the Model 50's Packard engines were replaced by air-cooled Pratt and Whitney Hornets, the aircraft being redesignated XPB–2. Only the one Boeing Model 50 was built.

The B & W, or Boeing and Westervelt. This small biplane, built by local carpenters and seamstresses in 1916, was the first Boeing.

The three-seat Boeing B–1, which was prevented from realizing its full commercial potential due to competition from cheap war-surplus aircraft.

Although there were to be no more Boeing flying boats for the next few years, apart from a relaunch of the Model 50 in 1929 as the Model 204 for commercial use, Boeing had in fact come into existence as a hydro-aeroplane manufacturer. Founded as Boeing and Westervelt, or B&W, in 1916 by Bill Boeing and Conrad Westervelt, the company's first design was known simply as the B&W, an attractive small biplane seaplane, built by local carpenters while the fabric on the wings had been sewn by local seamstresses. At least two of these aeroplanes were built for the New Zealand Government. Westervelt was recalled to the United States Navy in 1917, but Boeing continued the business after his partner's departure, building a refined B&W as the Boeing Model C and selling fifty of these to the United States Navy as trainers, while the United States Army Air Corps obtained two landplane variants. Later, the company built some HS–2L flying boats under licence from Curtiss, to meet the USN's terrific wartime appetite for aircraft, before returning to their own designs with the 50-foot span, B–1, a 90 mph three-seat flying boat with a 400-mile range and built entirely of wood. However, by this time the war had ended and the sale of surplus military and naval aircraft at 'knock down' prices prevented the B–1 from realizing its market potential and nearly forced Boeing out of the aircraft business. The company managed to survive by undertaking sub-contract work for other manufacturers and then by building landplane and carrier-borne fighters for both the USAAC and the

USN, though some of these, the carrier-borne FB–3 and FB–4 fighters, were also on rare occasions found fitted with floats.

In Britain, Porte's patents were by this time in the possession of the Gosport Aviation Company, itself on the verge of collapse, although Short Brothers were building a Porte-type flying boat as their 113-foot span Cromarty. Less fortunate was the latest in the line of Felixstowe designs, the triplane Fury flying boat, which crashed during take-off, killing its pilot after the hull started to porpoise at 27 knots. Porpoising was a problem with flying boats and amphibians, and while hull design had a bearing on the problem, it was most usually a question of inexperienced pilots not pulling the control column back – in the reverse of landplane practice – allowing the aircraft to bounce, burying its nose in the waves. Another factor which contributed to porpoising was the construction of aircraft with the centre of gravity too far forward.

On both sides of the North Atlantic, work on shipboard seaplane development continued after the end of World War I. Small flying boats, such as the AB–3, had been adapted for shipboard use during the war years, with the cruiser, USS *North Carolina*, the first American warship to operate aircraft on a regular operational basis, launching the AB–3 from a steam catapult for the first time on 12th July, 1916. A second cruiser, the USS *Huntingdon*, was also fitted with a steam catapult to operate two Curtiss floatplanes in July, 1917. However, these stern-mounted catapults required the warships to stop in order to launch and recover their aircraft, and, for this reason, both cruisers lost their aircraft in late 1917. An improved compressed air turntable catapult was developed and tested in 1921, initially fitted to the new Maryland-class battleships to enable these ships to carry a spotter aircraft for their 16-inch guns. Routine operations started from the USS *Maryland* herself on 24th May, 1922, using a Vought VE–7 floatplane.

The VE–7 had originally been developed in 1918 as a trainer for the USAAC, but the United States Navy also acquired twenty VE–7 – a single-float biplane with two seats and a 180-hp Wright-Hispano engine – introducing these from 27th May, 1920, onwards. Subsequently, some sixty VE–7s were built by Vought and another seventy by the Naval Aircraft Factory, followed in 1920 by the Wright-Hispano E–3 powered VE–9. The VE–7 eventually gave way to another Vought aircraft, the UO–1, essentially a VE–7 with a 200-hp Lawrence engine and certain aerodynamic modifications, but with the same large central float as the VE–7 series. Usually, American battleships operated three of these aircraft and cruisers operated two,

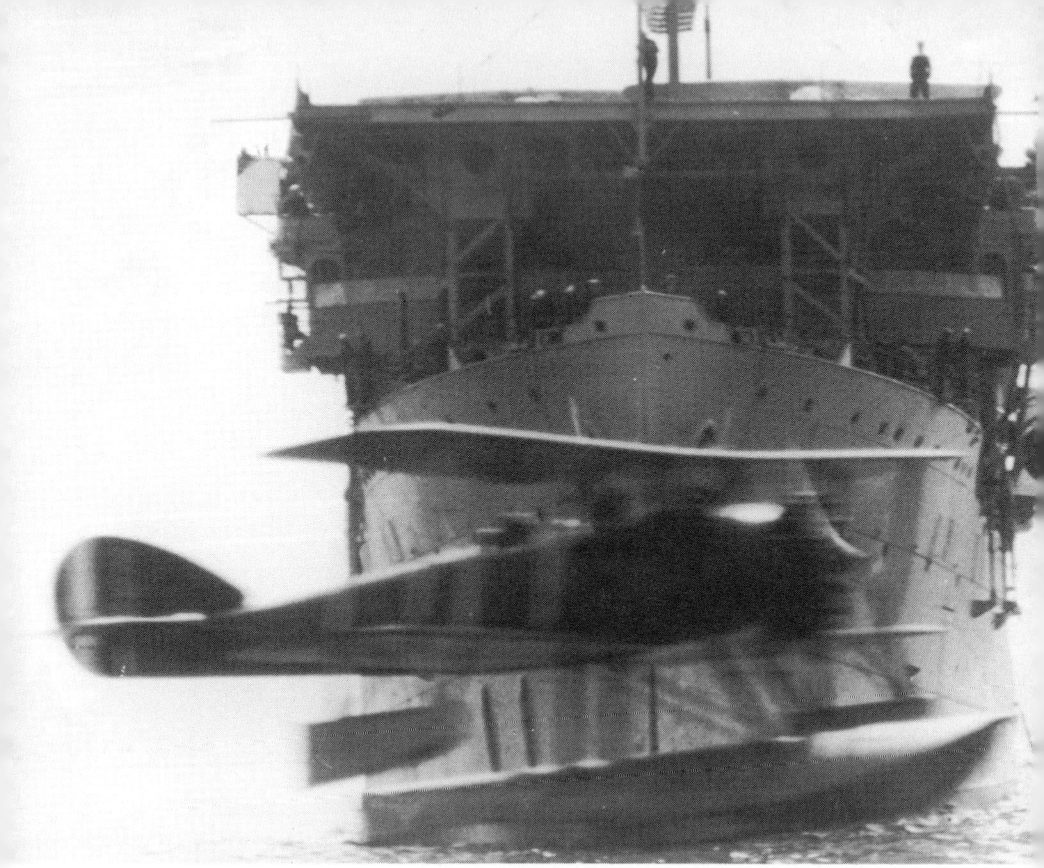

One of the post-World War I fleet reconnaissance aircraft, usually operated from battleships, was the Vought UO–1, seen here flying past the bows of USS *Langley* in 1924.

with the first operational deployment on a cruiser being aboard the USS *Tennessee* in June, 1924. A landplane fighter version of the UO–1, the FU–1, was also occasionally to be found operating on floats.

Other manufacturers also built seaplanes for the USN at this time, including Aeromarine, who in 1922 introduced their AS–1 and AS–2 twin-float seaplanes powered by single 300-hp Hispano engines. Martin built the first American all-metal aircraft, the MO–1, a three-seat spotter biplane seaplane powered by a 350-hp Curtiss D–12 engine, which was also deployed aboard battleships and cruisers. It was a 53-foot span aircraft of unprepossessing design.

The USN was also operating Douglas P2D–1 seaplanes at its base in the Panama Canal Zone, Coco Solo, and it was for operations in this region too that the USAAC maintained a strong interest in seaplanes and amphibians during the 1920s. Grover Loening, the former associate of the Wrights, proposed to the USAAC in 1923 that he should build an amphibian based on the de Havilland DH.4 bomber, which had been built in considerable numbers in the United States and was also, in fact, one of the first transport aircraft in its modified civil landplane form. A single-engined biplane of conventional design, the prototype Loening amphibian first flew on 9th June, 1924, and was designated XCOA–1. Unfortunately, the aircraft crashed during its first day of test flights, but in spite of this, a batch of ten production COA–1s, powered by Liberty engines, were delivered from January, 1925, onwards. The wheels on these aircraft folded into the hull. A further fifteen OA–1A, nine OA–1B and ten OA–1C amphibians followed, all being built at Loening's factory in New York. These were operated by the USAAC on patrol, survey and exploration duties. A development of the series was the XOA–2, delivered on 3rd July, 1929, and that was followed by eight production OA–2s, although these were built by Keystone after that company acquired Loening during the late 1920s.

Curtiss, meanwhile, had developed a number of landplane triplane and biplane racing aircraft for the United States Navy, including the R–6, a version of which, the CR–3, was developed as a twin-float seaplane for the 1923 Schneider Trophy seaplane race at Cowes, described in the previous chapter. The Naval Aircraft Factory married the R–6 fuselage to the HS–2 wings, with a 400-hp Liberty engine, to produce the PT–1 and PT–2 twin-seat torpedo-bomber seaplanes. Fifteen PT–1s and eighteen PT–2s were built between August, 1921, and June, 1922.

Using its eighteen PT–2s, USN Torpedo and Bombing Squadron 1 conducted a massive torpedo-dropping practice on 27th September, 1922, against the USS *Arkansas,* scoring eight hits with Mk.VII torpedoes dropped from the 96-mph aircraft. The USN's experiments with floatplanes at this time also included tests with Dutch Fokker FT–1 twin-seat low-wing monoplane seaplanes, which also used the 400-hp Liberty engine, and British single-seat Blackburn Swifts, although these were not strictly seaplanes but landplanes with 'releasable' wheels, emergency flotation gear and folding wings. However, it was another American design that emulated the production success of the UO–1 and the PT series, the 100-mph Douglas DT–2 two-seat biplane seaplane. The USN obtained thirty-nine DT–2s with 420-hp Liberty engines from Donald Douglas and a small number from Glenn Martin. Dayton–Wright also built a number of DT2s and so too did the Naval Aircraft Factory, while a number of DT–2s were uprated in 1923 using 525-hp Wright T–2s and redesignated DT–4s or, with Wright T–2Bs, were designated DT–5s.

A DT–2 could carry a 1,835-lb torpedo and all DTs were easily convertible from landplane to seaplane, or vice versa. After entering service with squadron VT–2 at San Diego in December, 1922, the DT–2 and its uprated developments saw service at Pearl Harbor and Hampton Roads. The DT–2 series were also exported to Norway and to Peru.

Following BuAer's practice of designing aircraft for construction by private manufacturers, Curtiss and Martin tendered to build an aircraft designated as the CS–2 and SC–1 respectively, but the Curtiss bid of $32,000 was dramatically undercut by Martin's bid of $20,000 per aircraft, allowing the latter manufacturer to build thirty-five three-seat Martin SC–1s between February and August, 1925, and after that, forty SC–2s. Boeing also built a BuAer design, the TB–1, which first flew in May, 1927.

Both the British and American governments attempted to introduce amphibians to their respective armed forces. In Britain, the role fulfilled by BuAer for the United States Navy was undertaken for the RAF, which at this time controlled all British service aviation, by the Air Ministry's Advisory Committee for Aeronautics. In 1920, the Advisory Committee organized a competition for an amphibian, with the flying test at Martlesham, near Felixstowe, proving to be an 'hilarious affair' according to one commentator. Supermarine offered one of their Channel flying boats, redesigned to incorporate an uprated engine and a folding undercarriage, while Fairey produced a modified World War I IIIC, and Vickers a Viking III flying boat in amphibian form. Although the Air Ministry was reputedly disappointed at the lack of any radical advance in design, they were also satisfied that a 'considerable advance had been obtained'. Vickers won the first prize of £10,000, leaving Supermarine to win the second prize of £8,000 and Fairey the third prize of just £2,000. Undismayed, Dick Fairey made much of the veteran IIIC in his Press advertising:

> She was delivered to the Royal Naval Air Service in April, 1917; was still in service after the Armistice; bought back in May, 1919, as a 'disposal' machine; used for experimental work during the summer of 1919; flown in the Schneider Trophy race in September, 1919, and was the only competitor to return intact under her own power; was used with experimental amphibian undercarriages during the early part of 1920, and has survived nine different engines. Her performance exceeds that of the best modern flying boats.

Clearly, a worthwhile achievement, showing the strength of the airframe, but surely there can have been few instances of someone making so much of coming third, and last! Fairey also forgot to mention that on the day of the trials his flight crew were first of all violently airsick and then, on landing, seasick!

On the Isle of Wight, Sam Saunders had designed his new Kittiwake amphibian, with the help of members of both Porte's and Sopwith's design team, but too late for the Air Ministry competition. However, the Kittiwake might not have done so well had it been in time to compete: it had suffered an accident during its test programme after the upper-wing leading edge had been blown off, there were numerous engine failures and there were problems with the ailerons. The aircraft was eventually written off in an accident while being flown by an Air Ministry pilot.

Fairey, Supermarine and a small manufacturer called Parnall were soon building amphibians for the Air Ministry. The Parnall Puffin suffered difficulties with the hydrodynamics of its single float, which did not run cleanly, so more chines were added to the float and it was lengthened, but this resulted in problems with the aircraft's aerodynamics. The Parnall Pintail amphibian also suffered problems that made longer floats necessary, but on the third aircraft the wheels were mounted on the inside edges of the two floats.

Fairey was having problems in the post-war recession in the aircraft industry, and in spite of a number of orders from overseas customers for IIID landplanes, the company was reduced to building bodies for Daimler motor cars in order to stay in business!

If Fairey had financial difficulties, Vickers was having other problems with its Viking IV amphibian. Sir Ross Smith and his brother, Sir Keith, planned to fly around the world in this aircraft, but during familiarization flights at Brooklands, the Vickers factory airfield at Weybridge, the aircraft developed a spin, and in spite of what appeared to be a recovery to observers on the ground, the aircraft spun out of the sky from 1,500 feet, killing Sir Ross Smith and the Vickers' pilot, J. M. Bennett.

However, such a set-back did not prevent Vickers from selling Viking IVs to the French Navy's air arm, the Aéronavale; to the Dutch for use in the Netherlands East Indies; to the Argentine Navy; to the Royal Canadian Air Force; to the Royal Air Force; and, just a single aircraft for evaluation in each case, to both the USSR and the United States Navy. Most of these aircraft were powered by Napier Lion engines, but the Canadian examples used Rolls-Royce engines.

The USN was having rather greater success with its amphibians at this time. As already mentioned, both the USN and the USAAC had obtained amphibians from Grover Loening, with the USN version of his aircraft being designated the OL–1. It differed from the Army versions in that it used a Packard engine instead of the original Liberty engine, while the fuselage was strengthened for operations from shipboard catapults. The first flight by an OL–1 was on 1st May, 1925 and five aircraft were ordered. These were followed by OL–3s and OL–4s for the USN and by the OL–5 for the United States Coast Guard service. The largest order came in August, 1926, for twenty-seven OL–6s with 475-hp Packard 2A–1500 engines, which were delivered to the United States Navy during the first half of 1927 and operated from their battleships. The United States Marine Corps also used these aircraft with their squadron in China at this time. The OL–6 was too big for the lifts aboard the American aircraft-carriers, and so a carrier-borne version, the OL–7, was developed with folding wings, but this aircraft then proved to be too heavy for carrier operations! Twenty examples of the next aircraft, the OL–8, were ordered in 1928, and these were fitted with wing-racks for four 116-lb bombs or up to ten 30-lb bombs. However, in spite of the rather aggressive specification, the aircraft still tended to be retained for the light transport, search and rescue, and utility roles for which the earlier aircraft in the series had been designed. After the acquisition of Loening by Keystone, one further aircraft in the OL series was built, the OL–9, and twenty-six of these were delivered to USN squadrons in 1931 and 1932.

Not all of the important flying boat, seaplane or amphibian manufacturers were British or American at this time. The success of the Italians in the Schneider Trophy event was evidence of this, while doubtless there would have been German participation as well had it not been for the fact that there was officially no such thing as a German aircraft industry, and German designers had to work in exile. At first, too, the Russian *émigré* Igor Sikorsky could not play a part until he could re-establish himself after the traumas of the Russian Revolution.

Undoubtedly, one of the most successful of the post-World War I flying boats, if not the most successful in financial terms, was the work of the German Professor Claude Dornier. Working at the Zeppelin Werke at Lindau, where he was chief designer, in 1918, Dornier had designed a small all-metal flying boat, the Gs.I, although by the time this aircraft was ready for its first flight, on 31st July, 1919, Germany had lost the war and the Allies insisted that the aircraft be destroyed in April, 1920. A development, the Gs.II, was banned.

Dornier avoided the problem of interruption to his work by the simple expedient of temporary exile in Italy, where the Società di Construzioni Meccaniche di Pisa, at the Marina di Pisa, started to build his designs under licence, enabling his Dornier Do.J Wal, or Whale, which incorporated much of his experience from the Gs.I and many of his Gs.II ideas, to make a successful first flight on 6th November, 1922. No less than 150 Wals were subsequently built in Italy, some by Piaggio, but most at Pisa, where the company changed its name to Construzioni Meccaniche Aeronautiche SA (CMASA) in 1930, following its acquisition by Fiat in 1929. A squat parasol monoplane flying boat, with two engines mounted in tandem in a single nacelle above the mainplane, the Wal used several different engines during its relatively long production life, which comprised some 300 aircraft built in several countries. Typically, two 600-hp BMW VI engines could provide a maximum speed of 140 mph and a range of up to 1,350 miles, even though it had a wingspan of just 76 feet and a 60 feet fuselage length. The Wal was one of the first flying boats to substitute hull-mounted seawings or sponsons, a Dornier innovation, for wingtip floats, providing additional stability, lower drag, some additional lift from what proved to be almost a sesquiplane effect, and additional fuel tanks. The first order for the Wal came from the Spanish Navy, which received six powered by 300-hp Hispano-Suiza engines in 1923. CASA in Spain eventually built some forty aircraft, while others were built under licence in the Netherlands by Aviolanda and in Japan by Kawasaki, and Dornier was able to start production at Friedrichshafen in 1932.

In spite of its civil registration, this is the military version of Dornier's highly successful Wal flying boat.

Inevitably, such a successful design fostered a number of developments, including the Dornier Do.15, an enlarged version of the Wal designed for the Luftwaffe, the new German Air Force, in 1933, and the Marine Fiat MF.5 built by CMASA in Italy during that same year. However, the most successful of the many Wal developments was the 'Ten Ton Wal', built by Dornier in Germany.

The first flight of the uprated Dornier Do.R Super Wal came on 30th September, 1926, using two 650-hp Rolls-Royce Condor engines, although most production Super Wals used four engines in two nacelles, again mounted above the wings. The first flight of a Super Wal, with four engines, came on 15th September, 1928. Relatively few of these aircraft were built, and most of the military versions came from CASA in Spain. It was usually known as the 'Ten Ton Wal'.

Two CMASA Wals were used by the Norwegian explorer, Roald Amundsen, in his attempt to reach the North Pole by air in 1925, and although he was unable to progress further than 87 degrees 43 minutes North on 21st May, this did establish a record for the most northerly point reached by an aeroplane of any kind. Amundsen sold one of his two Wals to Frank T. Courtney, who fitted two 450-hp Napier Lion engines to the aircraft for a transatlantic flight, although he had to abandon this after two failures, including one in which high winds forced him to make an emergency landing at La Corunna in Northern Spain. Courtney sold his aircraft to a third owner, Wolfgang von Grunau, who fitted the aircraft with new engines, two 600-hp BMWs, for a successful flight from Germany to Iceland, Greenland and New York, and then on to Chicago, between 20th and 26th August, 1930, marking the first east–west crossing of the North Atlantic. Known as the *Grönland Wal*, von Grunau's aircraft also made a round-the-world flight in 1932. A Spanish airman, a Major Franco, the elder brother of the later Caudillo, flew a Spanish CASA-built Wal *Plus Ultra*, fitted with Napier Lion engines, from Palos de Molguer, in Spain, to Buenos Aires, in Argentina – a total of 6,250 miles in 59½ flying hours – between 22nd January and 10th February, 1926.

Amundsen was not the only polar explorer to seek to use aircraft, although the American pioneer, Byrd, preferred landplanes. The veteran polar explorer Sir Ernest Shackleton obtained two Avro Baby biplane seaplanes for his expedition to the Antarctic, setting sail aboard the SS *Quest* from London in September, 1921. The small Baby seaplanes used single 80-hp Le Rhône rotary engines and instead of the normal wire bracing on the mainplane, tubular steel was used to ease maintenance. Whether or not such small aeroplanes, underpowered and with no protection from the elements for the pilot, would have performed well in the Antarctic is uncertain. In the event, the aircraft were never put to this most exacting test, for after the death of Shackleton aboard the SS *Quest* on 5th January, 1922, while the expedition was still on passage, the venture was cancelled and the aircraft returned aboard the ship to England.

It was clearly a time for exploration by aeroplane, however. While the planned round-the-world flight in the Vickers Viking IV of Sir Ross and Sir Keith had been cancelled after their tragedy, another round-the-world attempt was planned by a Major W. Blake, Captain Norman Macmillan and Lieutenant-Colonel L. E. Broone. However, this intrepid trio did not intend to use one aircraft, they proposed to take a D.H. 9 landplane from England to Calcutta, a Fairey III seaplane to Alaska, and a D.H. 9 for the flight across North America to New York, before taking a Canadian-built Felixstowe F.3 flying boat for the transatlantic flight. They actually started their journey, and all went well on the flight to Calcutta. However, while flying in their Fairey III, engine trouble forced them down in rough seas not far from Chittagong, then in India but now part of Bangladesh, and while taxiing in rough seas for some time, the aircraft developed a severe leak in the starboard or right float, so that eventually the aircraft sank, leaving the crew clinging to the other float until their rescue the following day – 24th August, 1922.

Another round-the-world flight using a Fairey IIID seaplane also came to grief when two Portuguese officers, Captains Sacadura Cabrul and Gago Coutinho, took two out of the three Fairey IIIDs obtained by

One of the most successful British seaplanes in the immediate post-war period was the Fairey III series.

the Portuguese Government in 1922. Starting in April, their first stage was to be a transatlantic flight crossing the central Atlantic from Lisbon to the Canaries and the Cape Verde Islands, then westwards to Brazil. However, mechanical failure and then the forced landing of one of the aircraft in heavy seas near St Paul's Rock, 500 miles off the Brazilian coast, brought an end to this attempt. Fortunately, a Portuguese warship was close by and able to render assistance to the hapless crew of the aircraft.

At this time, the old Isle of Wight boat-building firm of Saunders was mainly occupied building hulls for other manufacturers' flying boats, but, undaunted by the Kittiwake disaster, Saunders set about building another design of its own, the Valkyrie, for evaluation by the Air Ministry. However, this aircraft did not achieve the sort of success that manufacturers needed at this time, and, amongst other designs, Saunders found themselves building wooden hulls for the new Vickers Valentia flying boat.

Decidedly more successful was Reginald Mitchell's new Supermarine design, the Seagull, a biplane amphibian with a tractor propeller, a single inter-wing engine, and folding wings. Eight of these aircraft were sold to the Royal Air Force in 1923 after the service had completed its evaluation of the aircraft and its folding wings. The new Royal Australian Air Force also ordered the aircraft, while a version fitted with a pusher propeller was purchased by Spain as the Scarab. An enlarged version of this aircraft was the Swan, with two Rolls-Royce Eagle engines and forward folding wings, although the forward folding feature was soon abandoned as the extra weight could not be justified by the saving in stowage space. The Swan was hardly an elegant aircraft to look at, with a barge-type hull, although it seemed to perform well in the water and in the air, with Supermarine's test pilot, Henry Biard, able to show that it provided a useful advance in performance over the F.5 flying boat. Later versions were fitted with Napier Lion engines of 450 hp.

Two Supermarine Seagull amphibians took part in the 1922 King's Cup Air Race, open at this time to amphibians, seaplanes and flying boats, as well as landplanes. The 1922 course consisted of a route north from Martlesham Heath to Leith and Dumbarton in Scotland, before running south to Falmouth and east to the finishing point at Lee-on-Solent, on the coast between Southampton and Portsmouth. There were no compulsory stops, and competitors were permitted to alight whenever they wanted during the 950-mile flight, which started at 05.30. Some ten competitors were involved in the 1922 race, with the Fairey IIID seaplane joining the two Seagulls, one of which was flown by Henry Biard and the other by the Master of Semphill, who had been advising the Imperial Japanese Navy, at this time the purchaser of a number of different British seaplanes and flying boats. Although the landplanes started from Martlesham, the seaplanes and amphibians took off from nearby Felixstowe. Biard's propeller flew off during the northward leg, although he managed to land safely, while Semphill landed at Renfrew to mend a broken wire, and at Anglesey and Padstow for fuel, and was effectively out of the race as a result. The IIID suffered the ignominy of not crossing the finishing line, having to take off, fly across the finishing line and land again.

Nevertheless, already the Royal Air Force was finding that amphibians had distinct advantages over flying boats and landplanes, as the proponents of the amphibian would expect, although there were cost and weight penalties in such aircraft. Yet there were occasions when the extra cost, meaning fewer aircraft, and the extra weight, affecting the range, speed and warload of the amphibian, were acceptable penalties in return for the flexibility it bestowed upon the squadrons operating such aircraft. One of the roles which the Royal Air Force had to handle between the wars was the policing of many potential trouble spots in the Middle East and Africa, with air power proving to be economical in the use of manpower and often providing the main extent of British involvement, supporting locally raised forces. In Iraq during the early 1920s, the RAF found their two Vickers Viking V amphibians showed considerable advantages when compared with pure flying boats, and this encouraged Vickers to build an improved version, the Viking VII, although later this three-seat biplane was renamed the Vanellus. The Vanellus, with its pilot and two gunners, was the first aircraft in the Viking series to have rearward folding wings rather than the forward folding wings of the earlier aircraft – the folding wings were regarded as essential so that the aircraft could also operate as a fleet spotter, operating from British aircraft-carriers, even though at this stage many British maritime aircraft were still operating from cruisers and battleships. Some of the fleet spotter Vikings had fixed wings, but with the span reduced from 50 feet to 46 feet so that the aircraft would fit into the hangar lift well of the new aircraft-carrier HMS *Eagle*, allowing the flight deck to be cleared quickly as soon as the aircraft had landed, while aircraft could be readied for take-off with the least delay. During 1925, the Vanellus was tested in trials against the Supermarine Seagull II.

Nevertheless, the Air Ministry still required the longer range and the heavier lifting capability of the flying boat, and the first true post-war generation of flying boats was already under development. Shorts were still producing the F.5 as their duralumin metal-hulled S.2, based on experience with the small Short Cockle experimental aircraft. Supermarine was offering the Southampton, and Phoenix, who later became part of English Electric, adopting the parent company's name, were building two flying boats, the standard patrol-sized Kingston, and the smaller Ayr. The Ayr was a British attempt to overcome the fallibility of wingtip floats, which were frail, while the aircraft had to heel over considerably before a wingtip float could offer any stabilizing action, providing that the float didn't break first! The monoplane Ayr used a sponson, similar to that on the Dornier Wal, but, ironically, it never flew, since during take-off the sponson dug into the water and the aircraft rolled to starboard, digging the sponson further into the water and making take-off, or even a straight take-off run, impossible! This appears to have been a bad time for Phoenix, or the English Electric air department as it became, since the failure of the Ayr was followed by problems with the Kingston, the prototype of which sank in 1924 after striking driftwood. However, after salvage the prototype was rebuilt, and flight trials of this wooden-hulled machine, the first of five pre-production aircraft, showed it to be an improvement on earlier Phoenix designs. Yet, there were further problems, for during Air Ministry trials at Felixstowe on 28th April, 1925, the first production aircraft sank after the two inter-wing-mounted Napier Lion engines pulled out of their mountings, causing the wings to collapse: the pilot and observer escaped with slight injuries. These setbacks nearly caused the disbanding of the air department at English Electric.

Obviously, there was still much to learn about hull design and structural integrity. One partial solution, to the hull problems at least, was to use metal, and Supermarine soon followed Shorts' lead by introducing a duralumin hull on the Southampton.

Shorts, meanwhile, had flown their metal-hulled F.5, or S.2, for the first time on 5th January, 1925, with their chief test pilot, John Lankester Parker, at the controls. Although slightly nose-heavy in the air, the multi-chine metal-hulled aircraft threw up less spray than the wooden-hulled versions – helping to solve one of the greatest hazards of take-off and landing. The new structure was not to receive Air Ministry support without exhaustive trials. The aircraft arrived at Felixstowe for its tests in mid-March, and the aircraft, known as *Tin Five*, was subjected to extensive taxiing and flying tests and a series of tests to evaluate the effect of corrosion on the hull. On one occasion, the aircraft was stalled from a height of thirty feet into the sea, with the intention of springing its seams, but the aircraft survived even this harsh treatment.

A disappointment for Shorts was to follow. Although the company was invited to tender for a new all-metal flying boat, using two Rolls-Royce Condor engines, which the firm called the Singapore, a rival bid from Blackburn with the three-engined Iris flying boat won the Air Ministry competition. Oswald Short decided to press ahead and build his Singapore flying boat in the hope of gaining an order. Adding to his displeasure was the knowledge that Supermarine had been very lucky, receiving an Air Ministry contract for no less than six Southampton flying boats without the Air Ministry even examining a prototype! That said, the Air Ministry's unusual rashness proved to be a case of confidence well placed, for when Henry Biard took the prototype up for a test flight on 14th March, 1925, the aircraft performed so well that few modifications were needed, and these were mainly confined to a slight repositioning of the two inter-wing engines. One of the aircraft's more desirable features was the position of the pilot, well forward in the nose, with an excellent view clear of spray. Most Southamptons had a metal hull, but the one exception was the prototype, with a magnificent hull using cedar inside and mahogany on the outside, and even this aircraft showed a vast improvement in performance compared with earlier designs.

By this time, however, Shorts had accumulated some considerable experience on metal hulls. This was partly a result of building one of the early all-metal flying boats, the Short Cockle, a low-wing monoplane, which must also have been one of the smallest flying boats ever built, using just two Blackburn Tomtit engines of only 16 hp each. A small aircraft with a span of just 36 feet, the Cockle had a fuselage less than 25 feet long, while the aircraft was capable, in theory at least, of some 70 mph. The circular-framed fuselage was based on Shorts' experience with a landplane, the Silver Streak, and the framework was covered with stressed duralumin sheeting. It was in fact on the Cockle that Oswald Short introduced his S series of designation, making the Cockle S.1. However, at first all was not well, for on 18th September, 1924, the aircraft failed to become airborne in thirty minutes of effort on the River Medway, close to the Short works at Rochester. After the angle of incidence of the wings was changed, a test flight was finally achieved by John Lankester Parker on 7th November, 1924. Later, the tailplane was modified to provide improved directional stability, but the aircraft did not pass its Certificate of Airworthiness before passing to the Air Ministry at Felixstowe for trials. Two Bristol Cherub engines of 32 hp each were fitted in 1926, and the aircraft continued to fly until late in 1927, after which the Air Ministry used it for a series of trials into corrosion.

The Cockle's problems in becoming airborne illustrated one of the difficulties in operating the marine aeroplane, that of 'unsticking'. Too heavy a sea led to bouncing or porpoising, while too calm a surface could result in a suction effect at high speed during take-off, so that the aircraft would not 'unstick' and climb into the air. The experience of a good pilot was even more essential with these aircraft than with landplanes, while considerable development was still required in the shape of the hull and of a good planing profile. Shorts were amongst the first aircraft manufacturers to introduce a test tank for trials of different fuselage and hull forms, although their own finances limited the size of the tank to 250 feet at first.

Nevertheless, the trend towards metal hulls was a step forward. Metal hulls were stronger, less likely to leak than wooden hulls, and could not become waterlogged, although corrosion was seen as a possibility from the start and recognized as a danger, bearing in mind the very light metal structures used in aircraft. However, there were other problems with wooden aircraft which contributed to a growing appreciation of the merits of metal construction. The Royal Air Force was still operating Fairey IIID seaplanes in the Middle East, and the wooden floats of these biplanes were much appreciated by the local termites. So, rather than catering for their voracious appetites, the Air Ministry demanded metal floats! According to one historian, Richard Fairey wouldn't hear of metal floats on his aircraft, but during his absence from the works during a lengthy illness, a fellow director pressed ahead with construction of an all-metal Fairey III. The row that erupted when Fairey returned to work lasted for some weeks. That said, the all-metal III, which became the Fairey IIIF or Ferret, proved to be a very successful aircraft. One of the highlights of the IIIF's career with the RAF was the successful return flight by a formation of four IIIFs between Cairo and Cape Town between 30th March and 22nd May, 1927, under the command of the veteran pioneer, Air Commodore Charles Rumney Samson.

One of the most successful flying boats of the 1920s, and one with many formation long-distance flights to its credit, was designed and built in Italy, the Savoia-Marchetti S.M.55, designed by Ing. Alessandro Marchetti after he joined SIAI – Savoia de Societe Idrovolanti Alta Italia. This was one of the more unusual aircraft in a period during which aircraft design was far less settled onto firm principles than it is today. The S.M.55 was, in fact, not simply a twin-boom design but an aircraft with twin flying boat hulls.

Originally intended as a twin-engined torpedo bomber and mine-laying flying boat for the Regia Marina, the first flight of the prototype

S.M.55 was during August, 1924, with the aircraft using two 400-hp Lorraine-Dietrich engines built under licence in Italy by Isotta-Fraschini, and mounted above the mainplane in a single nacelle with one engine driving a tractor propeller and the other a pusher. The mainplane itself was extremely thick – it had to be, acting as it did as the connection between two substantial hulls, with the cockpit for two pilots being positioned in the leading edge of the wing midway between the hulls. Both hulls had a gunner's position, each with a 7.7-mm machine gun. Overall, the aircraft had a span of some 80 feet and was 54 feet in length, but the hulls were made of wood. In fact, both civil and military versions of the S.M.55 were built, with some 200 aircraft being produced for the Italian armed forces over a period of several years. Both the S.M.55 and its three-engined derivative, the S.M.66, remained in production from 1924 to 1940, which was a remarkable achievement for any aircraft, but especially for a design originating during the first half of the 1920s. In spite of this laudable achievement, however, the real claim to fame of the S.M.55 series lay in the prestige and propaganda flights of this aircraft under the Italian dictator Benito Mussolini during the 1920s and 1930s, even after it had gained fourteen records for speed, altitude and distance with varying payloads during 1926.

One of the first of many long-distance flights came in 1925, with the Marchese Francesco de Pinedo flying an S.M.55 from Rome to Tokyo. Later, in 1927, de Pinedo crossed the South Atlantic and flew north to the United States on an aircraft named the *Santa Maria*, although during what should have been a triumphant tour of the United States his aircraft burnt out in a refuelling incident in Arizona. A second *Santa Maria* was flown out to the United States, but *en route* the aircraft had to make a forced landing when it ran out of fuel and suffered the ignominy of being towed into harbour by a sailing schooner!

De Pinedo's flights were typical of the prestige-seeking and route-proving flights made by many pioneers during this period of the aeroplane's adolescence and eventual coming of age. However, while there was much of the pioneering spirit in de Pinedo's achievements, a rather less subtle flag-waving lay behind the efforts of his fellow countryman, the bombastic Italo Balbo, whose forte was the mounting of many long-distance mass flights, using the S.M.55 in its uprated S.M.55A version, with twin 750-hp Fiat or 800-hp Asso engines. For the longer flights, the gun positions were faired over to reduce drag, while of course guns were in themselves unnecessary weight on a record attempt. On 17th December, 1930, fourteen S.M.55As took off

One highly successful flying boat was the Blackburn Iris; this is a three-engined Iris V of 1932.

from Orbetello, the seaplane base north of Rome, and flew the 6,500 miles to Rio de Janeiro, arriving there on 6th January, 1931, although only ten aircraft finished the flight. Two of the aircraft crashed, killing five airmen, a third aircraft force-landed, collided with the vessel taking it in tow and then sank, while the fourth aircraft was damaged during take-off. The following year, S.M.55As took part in a 14,000-mile trans-Siberian flight.

Not all of the new flying boats of the period kept pace with the new trends in design or offered novelty, for metal hulls and monoplane configurations were far from being accepted, and there were occasions when it was an advantage for an aircraft not to be too far ahead of its competitors in design. The Blackburn Iris, for example, was a biplane with a wooden hull. The Iris was Blackburn's third flying boat, following the small N.1B and the ill-fated Schneider Trophy Pellet, which had failed even to make a successful first flight. Three 650-hp Rolls-Royce Condor engines powered the Iris prototype on its first flight in 1926, after which the production aircraft was designated the Iris II, and on this aircraft Blackburn allowed themselves to build a metal hull. Described by the popular Press of the day as a 'Giant Flying Ship', the Iris cost the awesome sum of £60,000, which scandalized many of the journalists and politicians of the day.

During August, 1927, the Blackburn Iris, Short Singapore,

Supermarine Southampton and the Saunders–Roe Valkyrie, were taken by the RAF on a Baltic Cruise, flying 3,000 miles to visit the Scandinavian capitals and the major Baltic ports. Both the Iris and the Singapore proved completely reliable during the long trip, with frequent take-offs and landings, and although the Iris with its deeper hull was the better aircraft on the water, the Singapore proved superior in the air. Both the Southampton and the Valkyrie suffered from engine trouble, but with the exception of a forced landing by the Valkyrie, which was towed into Königsberg for repair, the flights to and from the UK, and the circuit of the Baltic, were all completed without undue incident.

Alliott Verdon Roe had by this time severed his connection with the Avro company he had founded before World War I and, after a period on his own, had joined Saunders on the Isle of Wight, the Saunders concern becoming Saunders–Roe or, as in British aviation parlance, Saro.

The Southampton had been developed from the Swan amphibian as the RAF's first true post-war flying boat. Although the prototype Mark 1 used a wooden hull, and this was delivered to No. 480 Flight at Calshot, near Southampton, before eventually joining No.201 squadron in January, 1929, both the Mark II and Mark III, which were the main production versions, used duralumin hulls. Some seventy-eight Southamptons were built, serving with 201, 203, 204, 205 and 210 squadrons in the United Kingdom, Iraq, and Singapore. Two inter-wing mounted 500-hp Napier Lion engines powered the Mark II, and all aircraft of this type used a triple fin and had tandem open cockpits. Capable of lifting 1,100 lbs in bombs or torpedoes, the aircraft had a maximum speed of 108 mph and a range of 770 miles. Ever anxious for improvement, Supermarine's talented designer, Reginald Mitchell, developed an improved Mark IV Southampton with twin 525-hp Rolls-Royce Kestrel engines, although this aircraft was redesignated as the Scapa in October, 1933. The Scapa retained the fabric wings and duralumin hull of the later Southamptons, but this aircraft had a simplified twin-fin tailplane and side-by-side seating for the pilots. Armament was the same on the Southampton and the Scapa, with bow and two amidships 0.303 Lewis guns. However, the Scapa offered 140 mph and a range of 1,100 miles. Only fifteen Scapas were built, serving initially with 202 squadron, and later with 204 and 240 squadrons in Malta and the United Kingdom. A final development of the Southampton was the Stranraer of 1935, with two 875-hp Bristol Pegasus radial engines faired into the upper wing. A tail gunner's position was added to this version, although, unlike the pilot's enclosed cockpit, the gun positions remained open to the elements. Just seven Stranraers were built, in addition to the prototype, for the RAF, but these aircraft served with four squadrons at different times, and a further forty were built by Canadian Vickers for the Royal Canadian Air Force, who kept the aircraft in service until 1943, three years after it was withdrawn from RAF service.

The Iris did not enjoy the same success. The second aircraft, the prototype Iris III, did not fly until 1929 and was delivered to the RAF that November. In January, 1930, the first and only Iris squadron, No.209, was formed with four aircraft. The Iris III, with 0.303 Lewis guns in the nose and tail, and amidships, was powered by three 825-hp Rolls-Royce Buzzard engines, giving it a maximum speed of 120 mph, a range of 470 miles, and provision for a 2,000-lb warload. The Iris remained with No.209 until 1934, when it was replaced by another Blackburn aircraft, the Perth, which first flew in 1933 and used uprated Buzzards for a maximum speed of 132 mph and a range of 1,300 miles. In common with the Stranraer, the Perth introduced enclosed cockpits for the pilots, while leaving the gun positions open. One feature of the Perth was the ability of the bow positions to fire, in addition to a 0.303 Lewis gun, a 37-mm gun capable of firing up to a hundred 1½-lb shells per minute! A more advanced Blackburn design, which achieved less success than it deserved, was the Sydney with three 535-lb Rolls-Royce engines faired into the monoplane wing, and which first flew in 1932.

Both Saro and Shorts were busy at this time with other projects, including the Saro A.3 Valkyrie and a competitor to the Seagull, the Medina amphibian. While Saro was also building floats for seaplanes for another small manufacturer, Parnall, Shorts had been experimenting with another small flying boat, the Mussel, with a cantilever monoplane configuration. The full-sized Short Singapore flying boat had made its first flight at Rochester in August, 1926, at the hands of John Lankester Parker, although the flight lasted just two minutes after an engine cowling became detached and fell off. Unfortunately, other modifications were also required, and the aircraft next flew on 6th October, 1926. The strain of developing the Singapore, and the limited number of orders for aircraft, meant that Shorts developed a cash flow problem during this period. To overcome this, the factory had to turn to building bus bodies for a time.

Meanwhile, the Supermarine Southampton had been going from strength to strength, and as a foretaste of its Baltic Cruise of 1927, the aircraft had flown across the Irish Sea in 1925, landing at various ports to demonstrate the ability of the aircraft to operate from improvised landing spots in the event of a crisis. During August, 1926, two Southamptons flew from Felixstowe to Egypt and back, a distance of 7,000 miles. In October, 1927, after the Baltic cruise triumph, four Southampton flying boats left for a twelve-month tour throughout the Far East, under the command of Group Captain Cave-Brown-Cave, to 'show the flag' by visiting major ports in the traditional manner of warships. All in all, the aircraft each flew some 27,000 miles during their tour, which lasted until December, 1928, rather further than Sir Alan Cobham's famous round-Africa flight. Some idea of the problems of operating flying boats, particularly when away from prepared and well-run bases, can be gained from an account by Squadron Leader G. E. Livock, RAF DFC, a pilot in one of the Far East flight Southamptons.

> As our safe range was only 500 miles, we had to alight 29 times for fuel between England and Singapore, partly because we had to fly around the coast of India, instead of crossing direct.
>
> Refuelling was one of our main headaches. As we lay at moorings – often in swift flowing rivers or rough water – the petrol had to be brought out in local boats handled often by native crews who knew nothing of flying boats and their fragility. We never knew what to expect, and usually had to handle the drums or tins ourselves and hand pump the petrol to the top wing tanks – arduous work after several hours flying in the heat of the tropics.

They arrived at Singapore and remained there until mid-May, continuing to Australia in late May and visiting Hong Kong afterwards.

Sir Alan Cobham had obtained the loan of the Rolls-Royce Condor-engined Short Singapore prototype for a survey flight of Africa, seeking landing places both for the Royal Air Force and for Imperial Airways. The aircraft received a civil registration for its survey flight, becoming G-EBUP. When Cobham left Southampton with the aircraft on 17th November, 1927, he flew via London, an indirect route for publicity purposes, to Malta, eventually reaching St Paul's Bay, near the north of the island, as dusk fell on 24th November, to land alongside the battleship, HMS *Queen Elizabeth*. The next day he took off for the RAF seaplane station at Kalafrana on the south side of the island, alighting safely in a less well-sheltered anchorage, only to find that, after dark, a large wave tore off the starboard, or right, float, and the aircraft was only saved by the crew climbing aboard and standing on the port wing to lift the starboard wings clear of the water. Before the aircraft could be safely slipped, high seas and high winds had

Sir Alan Cobham borrowed this RAF Short Singapore, which was given a civil registration for his flight through Africa.

damaged thc other float and then the lower wing. It was not until 21st January, 1928, that Cobham's aircraft was repaired from spares sent out from England, so that he could continue his 23,000-mile flight, returning home in June to hand the aircraft back to the Royal Air Force. On receiving their flying boat back, the RAF soon fitted it with the new Rolls-Royce Buzzard engines, raising the maximum speed to 132 mph. In spite of the problems at Malta, more than anything else the arduous flight round Africa proved the value of a metal hull.

A development of the Singapore, a three-engined flying boat known as the Short S.8 Calcutta, was built for Imperial Airways, and a naval derivative of this aircraft, a long-range maritime reconnaissance flying boat, the S.15, was ordered in 1930 by the Imperial Japanese Navy, with the aircraft being dismantled and shipped to Japan after a successful test flight in Britain. Assembly in Japan was by Short's licensee, Kawanishi, who were also Rolls-Royce's licensee, and subsequently the design was used as the basis of a new series of Japanese flying boats. A sole Calcutta was built for the French Aéronavale, before Breguet built four aircraft under licence and then used the design as the basis for its Br.521 Bizerte flying boat. Even the Royal Air Force obtained a Calcutta development, the S.8/8 Rangoon, with four built and operated in Iraq before eventually being transferred back to the UK.

The Breguet Br.521 Bizerte showed its Calcutta ancestry, with its three inter-wing engines and the same tailplane configuration, although the engines were more powerful 900-hp Gnome-Rhône radials with a maximum speed of 160 mph and a range of 1,245 miles. A civil version of the Bizerte was the Saigon. First flown on 11th September, 1933, at Le Havre, the Bizerte was allocated to the Aéronavale Escadrille, or squadron, E.1, at Cherbourg. The three production aircraft which followed the prototype included two in which the cockpit was extended forward to the tip of the nose, enclosing the original position of the bow turret. This firing position was replaced by two guns mounted in blisters aft of the cockpit, with forward firing 7.5-mm Darne machine guns. Two more machine guns fired through hatches amidships and a fifth gun was positioned in the tail. Economic difficulties meant that the Bizerte could only be built in limited numbers, joining Escadrille E.2 and E.3 in late 1935, and only twelve aircraft had been built by the end of 1936, and twenty-eight by late 1938, with only three more built before the fall of France in 1940.

The French Aéronavale badly needed aircraft with the performance of the Bizerte, having been largely dependent upon obsolete designs. A French mainstay of the Aéronavale at the time was the CAMS 37, a wooden biplane flying boat with inter-wing engines first flown in 1926, and the twin-engined wooden biplane CAMS 55, which followed in 1929. Some more modern designs were to follow, including the Lioré 501 multi-purpose amphibian of 1931, a high-wing monoplane with a single engine driving a pusher propeller mounted above the mainplane. A rather sleeker design was the single-engined Potez 452 of 1932, a small flying boat with the engine faired into the monoplane mainplane. Sixteen of these were built – a not inconsiderable number for a French design at that difficult time.

For the time being, the biplane continued to rule supreme whenever an aircraft was needed for either carrier operation or for operations from battleships and cruisers, although interest was growing in the monoplane for landplanes and for 'shore-based' maritime aircraft. This was true on both sides of the Atlantic, the lower drag of the monoplane being disregarded for the more compact shape and better lift of the biplane with its relatively low wing loadings. In the United States, one of the classic spotter aircraft was being built by Chance–Vought, the O2U–1 Corsair biplane, using a Pratt and Whitney Wasp radial engine and, in the usual manner of American seaplanes, balancing on a single float. First flown on 2nd November, 1926, as a landplane, one of the first seaplane versions of this aircraft set a record in its class for speed and altitude in April, 1927, while even the standard machine could manage 147 mph and an altitude of 18,700 feet – not too far behind the speed range of contemporary fighter aircraft. By February, 1930, some 130 aircraft of this type had been built, with the most numerous variants being the O2U–3 (eighty) and O2U–4 (forty-two). Aircraft of this type served aboard every United States Navy battleship and cruiser. Although most Corsairs were landplanes or seaplanes, a small number received amphibian floats built by the new Grumman concern, which was also building floats at this time for a Brewster seaplane, retaining the concept of a single large seaplane float, into which the amphibian wheels could retract.

A far less conventional ship-borne, or as some purists would have it, boat-borne, aircraft of this time was the Parnall Peto, a diminutive biplane seaplane intended to be, and which in fact was, launched from a submarine. The idea of a submarine-launched aircraft was not new, for some thought had been devoted to the concept before the outbreak of World War I. However, a suitable aircraft and a submarine class of adequate size had not been forthcoming at this early date. A submarine 'aircraft-carrier' first appeared with the Royal Navy's 'M' class of three submarines. These had been completed with each vessel having a 12-inch gun from a Majestic class battleship, which the submarine was supposed to fire as it surfaced briefly close to a suitable target – a form of attack favoured by several navies, including the French, who also dabbled with such craft, calling them 'corsair' submarines. The idea had novelty, but little else to commend it, and the British Admiralty eventually took the understandable view that any target worth a 12-inch shell would be well worth a torpedo, which could be discharged with far less risk to the submarine or its crew. While the 'M' class was withdrawn from service, one of the submarines, M2, was converted to carry a small seaplane and recommissioned. A single Parnall Peto biplane was designed and built for operations from M2, on which a hangar door folded flat after the submarine had surfaced, so that a catapult launcher could be quickly assembled over the door and the deck while the aircraft was brought out of its small hangar, which was just eight feet wide, and its wings unfolded. The whole operation took but a few minutes. On landing in the sea after a flight, the Peto could be picked up by a derrick fitted to the conning-tower and positioned on the catapult rail, so that, with wings folded, it could be pushed back into the hangar, and the catapult folded away.

Much was made of the safety precautions incorporated into this procedure to avoid the risk of the submarine being sunk. Everything was done to minimize the risk of the hangar door being open at the same time as the internal hatch between the hangar and the conning-

The launching and hangar arrangements of the Parnall Peto aboard the submarine M–2 can be clearly seen here.

tower, although there was no physical safety mechanism which would prevent both doors from being open at the same time.

The small, under-powered, two-seat Peto was fitted with a radio for reconnaissance duties, but the aircraft was unarmed. True or false, the story current at the time was that the observer had to leave his flying boots behind and wear plimsolls if the pilot wished to carry any additional equipment! The aircraft had the reputation of handling badly on landing in the open sea. There were compensations, however, with the crew receiving double danger money, flying pay and submarine pay, in addition to their normal service pay.

Trials started in 1927, with a flight by Lieutenants C. W. Byron and

C. Keighley, RN. In spite of the aircraft's limitations, several years of successful flights followed, before tragedy struck. On 26th January, 1932, the M2 dived off Portland, in the western approaches to the English Channel, never to surface again. Divers located the submarine, but some sixty officers and men had died and attempts to raise the wreck failed despite the efforts of experienced Royal Navy salvage teams, who during the summer of that year very nearly brought the vessel to the surface, but were defeated on several occasions by bad weather, before eventually salvage attempts were abandoned. It is generally thought that the M2 shipped a heavy sea through the hangar door and into the conning-tower hatch, swamping the submarine – certainly this would be the most logical explanation for such a sudden disappearance.

The French corsair submarine, *Surcouf*, was designed from the start as an aircraft-carrying submarine, with a hangar aft of her conning-tower.

In addition to reconnaissance, rescue operations had also become a part of the role of the flying boat and seaplane by this time. A lower standard of prosperity and the relative absence of paid holidays meant that there were fewer instances of yachtsmen and holidaymakers getting themselves into difficulties, but the poor reliability of the aeroplane at the time also meant that there was a need for adequate search and rescue services. One search and rescue operation which was far from routine, however, arose when the Italian explorer, General Nobile, went missing in his airship, *Italie*, in late May, 1928, in an attempt to fly over the North Pole. Seaplanes and flying boats from Italy, Norway, Sweden, France and Russia joined in the search, using Spitsbergen, to the north of Norway, as the base for the search operation. In fact, a small de Havilland Moth landplane eventually found Nobile. The flying boat itself was still far from reliable, and during the search, the Norwegian explorer Amundsen and two colleagues were killed when their Latham flying boat crashed, with the only item of wreckage recovered being a float, which was washed ashore near Tromsö.

In the United States, and indeed elsewhere, as we have already seen with the development of the Southampton, existing designs were exploited to the full. The Chance–Vought O2U–1 had led to a number of developments including the O3U–1, with improved performance, which entered service in July, 1930, although this variant was more usually operated as a landplane. A need for a lighter observation aircraft appeared at this time, as the Corsair series proved to be too heavy for the catapults on the Omaha-class cruisers, which needed an

A monoplane development of the Iris was the Blackburn Sydney, a design some years ahead of its time in 1930.

aircraft of less than 5,450 lbs all-up weight. This led the Bureau of Aeronautics to evaluate prototype aircraft, including the Keystone XOK–1 and the Berliner–Joyce XOJ–1, both of which used 400-hp engines – the Wright Whirlwind and the Pratt and Whitney Wasp, respectively. Eventually the OJ–2 was ordered into production as a development of the Berliner-Joyce XOJ–1, entering service with the United States Navy from 1933 onwards, although in relatively low numbers.

Change was taking place by this time, as the enhanced performance of the landplane for torpedo and bomber roles meant that the seaplane was beginning to lose its attraction, being unable to overcome the drag of its floats to compete with landplanes for these roles. The change was most marked in those navies which had adopted the aircraft-carrier, and could position landplanes wherever they were needed. The irony was that this decline of the seaplane set in before the seaplane had given of its best, but the landplane was also set to improve, and from this time on, the seaplane became increasingly the preserve of the battleship or cruiser, which required an aircraft capable of making a water landing.

By contrast, the flying boat and amphibian were approaching the peak of their utility, offering improved performance and a solution to the problems of maintaining adequate maritime reconnaissance and of providing transport and communications to the more remote areas of the world. For the Royal Air Force, the flying boat was the answer to the need to police an empire, for the United States Navy, the flying boat was the means of maintaining contact with, and ensuring that there would be no disruption to the shipping routes to Hawaii, while also helping to police the Panama Canal Zone. The single long hop from the west coast of the United States to Hawaii meant that range was even more important to the USN than it was to the RAF, which had to provide air cover over a substantial part of the earth's surface.

However, by the late 1920s, the United States Navy required flying boats in substantial numbers not only for operations to and from, and around, Hawaii and the Panama Canal Zone but also to Alaska and to the Philippines, over which the United States had certain commitments not so very different from many of those bestowed upon the Royal Air Force in the Middle East, and elsewhere. While BuAer was still handling aircraft design and placing contracts with private manufacturers, the decision was taken to order a USN design from Consolidated, a prototype known as the XPY–1, using two 450-hp Pratt and Whitney engines mounted on struts below the large monoplane wing and above the fuselage 'parasol' fashion. Ordered in February, 1928, the aircraft featured a duralumin hull, metal-framed wings, although they were still fabric-covered, and a twin fin. After early trials, a third Wasp engine was added above the mainplane to improve performance. Under BuAer's system, construction of the prototype, no matter how successful, was no guarantee of a production contract, and when Glenn Martin undercut Consolidated he received the contract to build nine aircraft, designated the P3M–1. These used twin Wasps initially, but from the fourth aircraft onwards two 525-hp Pratt and Whitney Hornet radial engines were used, enclosed in new ring cowls developed by the National Advisory Committee on Aeronautics. These aircraft were designated the P3M–2, and had enclosed cockpits for the pilots. Later, all the Martin aircraft were brought up to this standard.

The move to the monoplane had already been made by the USN, but there was one more patrol flying boat of biplane configuration to come, the six-seat Hall XP2H–1, with four 600-hp Curtiss Conqueror engines in tandem pairs between the wings. Capable of lifting a 2,000-lb warload, the XP2H–1 had Browning machine guns in nose, dorsal, waist and tail positions – the first tail positions in an American aircraft. Consolidated continued with flying-boat development and a contract was awarded to the company for an XP2Y prototype, which first flew on 26th March, 1932, with three engines, though the aircraft was modified and flew with just two engines on 18th May, after which

A more conventional biplane development of the Iris was the Blackburn Perth, first flown in 1933.

Consolidated was awarded a production contract for twenty-three twin-engined P2Y–1s, the first of which reached VP–10 at Norfolk, Virginia, on 1st February, 1933. The P2Y–1 soon showed that it was more than capable of meeting the USN's expectations for it in terms of endurance and reliability on long over-water flights: aircraft of VP–5 flew non-stop over the 2,059 miles from Norfolk to Coco Solo in the Panama Canal Zone in September, 1933, taking 25 hours 20 minutes, while six aircraft of VP–10 flew non-stop from San Francisco to Pearl Harbor in January, 1934, covering 2,400 miles in 24 hours, 35 minutes. A development, the XP2Y–2, first flew in August, 1933, with the engines faired into the leading edge of the upper wing in what was by now a sesquiplane design. More than twenty P2Y–3s were delivered to the USN during the first half of 1935, using 700-hp Wright Cyclone engines, while the type was also sold to Colombia, Argentina and to Japan, where Kawanishi assembled the aircraft from kits as well. Doubtless this design helped to influence Japanese aircraft design, being built under licence as the Navy Type 99, or H5Y1 'Cherry' flying boat, even though the Japanese version was generally regarded as being under-powered.

A change in favour of the monoplane, enclosed cockpits and aircraft with longer range and greater reliability was already apparent as the maritime aeroplane entered the 1930s, probably its greatest decade.

5
BRIDGING THE OCEANS

As peace returned to Europe (it had never really left most of the rest of the world) a large number of surplus aircraft and surplus wartime-trained air and ground crew became available. For many for whom flying had been but a dream in 1914, it was an experience by 1918, while for others, who had worked as blacksmiths or had worked on farms or in factories, there had been an opportunity to learn and to understand the internal combustion engine. Air transport was not to be the sole beneficiary of this step forward, since road transport was also to enjoy a rebirth to the discomfort of the railways, but while road transport, both passenger and freight, was to expand rapidly, in the air the effort had more of the pioneering spirit, providing excitement and novelty, but uncertainty as well. Europe did not have a network of civil airfields, let alone landing or navigational aids. The early 'airliners' were in fact converted bombers, initially the de Havilland-designed Airco D.H.4 and D.H.9, and then later the Vickers Vimy and the giant French Farman bombers. These were all landplanes, and at first the new airlines did not seem to be interested in seaplanes or flying boats. This picture was to change as the British and French airlines pushed out towards their respective empires, and the Americans attempted to cover the Caribbean and then the Pacific.

The novelty of the aeroplane was still around, and indeed the aeroplane had itself been in existence for just over fifteen years by the time World War I ended. As a result of this novelty, no one really seemed to be sure whether the flying boat or the airship would be the ideal air transport for the longer over-water distances. At first too, the new airlines were on their own, and although before long subsidies were given to European airlines, British and American airlines had to stand on their own feet at first, before air mail contracts provided some assistance, and later some measure of subsidy was given to a new British national airline. Such help as there was from the British Government in the early days tended to be indirect, with the Royal Air Force exploring and surveying forty-three landing sites for flying boats between Cairo and the Cape, for example. However, that was of little use to the airlines struggling against subsidized French and Dutch competition on the cross-Channel routes, to the extent that by the winter of 1919–20, the British pioneer of air travel, Aircraft Transport and Travel, was bankrupt! Elsewhere that year, a young American student at Yale, Juan Trippe, born in 1899, saw potential for the flying boat in transatlantic air travel, inspired by the flights of the Curtiss NC–4. However, air travel was limited initially to operations between London and Paris, Brussels and Amsterdam, and for these routes one didn't need flying boats!

Of course, there were those who were quick off the mark – too quick in some instances. One of the most ambitious of the early post-war projects was that by the new Italian designer, Giovanni Caproni, building a 66-foot-long flying boat with three triplane planes, room for a hundred passengers, plus a pilot, two co-pilots and three engineers, who had to look after eight 400-hp Liberty engines. Intended for transatlantic flight, this ungainly aircraft was seriously underpowered, and on the second of its two flights, in 1919, crashed into Lake Maggiore, after which the entire project was abandoned. In complete contrast, a small airline, LARA or Ligne Aérienne du Roi Albert, operated a Levy flying boat in the Belgian Congo, carrying up to two passengers on a route between Leopoldville and Stanleyville, and in fact carrying just ninety-five passengers during its entire life of two years!

As early as 1919, a company called Aeromarine Airways used a converted Curtiss-built F–5L flying boat for irregular passenger and freight flights between Key West, off the Florida coast, and Havana in Cuba, although the operation folded in 1924 before the airline had time to introduce a planned Miami–Bimini–Nassau route. The great American hopeful, Rod Wannamaker, for whom the Curtiss–Porte America had been designed, had formed an airline in 1916, called the American Trans-Oceanic Company, which commenced operations on 24th February, 1920. Using a Curtiss H–16 which had been modified to carry ten passengers and was painted to look like a large fish, the plane flew between Miami and Bimini carrying sea anglers and other sport fishermen. This airline also collapsed after the H–16 made a heavy landing in 1921, presumably because its management did not feel that the financial returns justified the cost of repairing heavy hull damage.

So, the new airlines were mainly landplane operators, with the English Channel acting as a spur to relatively short distance air travel within Europe. At first, aircrew wore a variety of clothes or overalls, but in 1921, a British company, the Instone Airline, introduced the naval style of pilots' uniform which is now standard throughout most of the world, timing this to coincide with its move to the new aerodrome at Croydon, on the main road from London to Brighton.

However, the relative absence of airfields throughout much of the world was to be the making of the seaplane and flying boat, and to an extent the amphibian. When Finnair was established in 1923, its founder, Bruno Lucander, used a single Junkers F–13 all-metal single-engined seaplane, initially on charter flights, but then on scheduled flights between Helsinki and Tallinn in Estonia, and to Stockholm across the Baltic in Sweden, during 1924, although at first such services were only operated during the summer months. The utility of the aeroplane was apparent even at this early stage, with the winter operations consisting of emergency services within Finland. After an exceptionally severe winter in 1925–26, the Finnish Government was sufficiently impressed to provide loans for the development of Finnair. By 1928, Finnair was operating four F–13s and a Junkers Ju–24, but it was not until the early 1930s that the airline operated its first landplanes.

The Junkers F–13 seaplane was also operated by SCADTA, the German-owned predecessor of Avianca, the Colombian airline, to operate from the many rivers in that jungle-covered country.

Finnair was not the only airline to be born in 1923. Two other air services started. One of these was in the United States, where Juan Trippe and his friend, John Hamilton, acquired nine ex-USN Aeronca seaplanes for just $500 apiece, to operate a semi-regular air taxi service known as Long Island Airways, flying between New York and Long Island, carrying tourists to fashionable beach resorts. However, this airline also ceased operations after two years, partly due to the

Finnair started operations with this Junkers F–13 seaplane, which operated across the Baltic and within Finland.

condition of the aircraft and partly to the poor reliability of the pilots and mechanics, who tended to report for duty only when they chose to do so! A rather longer term operation was established in Britain, where the British Marine Air Navigation Company bought three Supermarine Sea Eagle amphibians. The Sea Eagle was a biplane with a single inter-wing engine driving a pusher propeller and a fuselage reminiscent of that of a cabin cruiser, and proved to be ideal for the services operated by the airline between Southampton and the Channel Islands of Jersey and Guernsey, and later Cherbourg in France: these routes continue today, with several flights daily, all operated by landplanes.

Operating aircraft at this time was a hazardous business, and one of the many crippling costs born by the unfinanced airlines was that of insurance. However, greater expertise in assessing insurance risks, and lower premiums, resulted from the formation, in 1924, of the British Aviation Insurance group. This coordinated and combined the efforts of the many companies which had dabbled in the insurance market, and which sometimes had charged premiums as high as 30 per cent!

However, the real interest in 1924 was the formation of a new Government-supported British airline, Imperial Airways, absorbing the operations of the Handley Page Transport Company, a subsidiary of the aircraft manufacturer, Instone Air Line, Daimler Airways and the flying boat operator, British Marine Air Navigation Company. The new airline was partially owned by the Government-sponsored British, Foreign and Colonial Corporation, which held half of the company's £1 million share capital, while a £1 million direct subsidy was payable for the first time in British aviation history, starting at £137,000 in the first year and reducing to £32,000 in the tenth year, on condition that the airline completed at least one million revenue miles annually. Pilots were, for the first time, required to have had a course of instruction on the aircraft they were to fly, and all pilots were to be members of the Royal Auxiliary Air Force or to have been members of the RAF. Two Government-appointed members joined the board, on which there were also members of the newly absorbed airlines.

The new airline got off to a bad start under the chairmanship of Sir Eric Geddes: the pilots went on strike! Imperial Airways claimed that a salary of between £750 and £850 per annum was an improvement over the pilots' private enterprise pay of £680, but the pilots claimed that Handley Page had paid £915, Instone £857 and Daimler £1,000. A month passed before the dispute was resolved, and only then largely because Imperial threatened to recruit new pilots, which would have been no difficult task during the post-World War I years. A salary of £750 was about five times the then average rate for a skilled worker.

However, the new airlines had a brief to develop air services to the Empire, exploration of new landing sites in the Far East and Africa for the long routes to South Africa and Australia (the latter being planned to operate through India and Singapore) being early priorities. Sir Alan Cobham's round-Africa flight in a Short Singapore flying boat is mentioned in the previous chapter. Most of his 1926 flight to Australia was in a de Havilland D.H.50 landplane converted to a seaplane. There had been plans to operate the aircraft as a landplane throughout the flight, except for the stages between India and Australia but these were reversed and the aircraft flew as a seaplane throughout, except for the flight between Darwin and Melbourne, when it was converted back to a landplane to fly over the Australian outback. So much for the chances of a landplane on what were to be the most important routes for Imperial Airways. Only the flying boat was to be able to operate over this route for some years, offering greater safety in the case of a forced landing and avoiding the cost of building runways.

Cobham, proponent of the flying boat and the seaplane, did not disguise the hazards of their operation, and writing of flying on the edge of the Timor Sea in New Guinea, he stated: 'The water is so clear, you cannot see its surface, and the crest of a wave running across its surface gives the impression of a bucket of water thrown onto glass. Alighting in the water is therefore very dangerous.' Obviously, even an experienced pilot could pull out too high and stall into the sea or, worse still, fly into the water!

Aircraft were needed for the new routes, and Imperial – which made a profit of £11,000 in 1926–27, but was still operating 92 per cent of its scheduled flights within Europe, with the only other route of any importance being the new service to Basra in the Persian Gulf – decided on a three-engined development of the Singapore, the Short S.8 Calcutta. This was almost an act of faith, for although the Singapore was flying at the time of the Calcutta being ordered, Cobham had still to make his African flight. Ordered in 1927, the first Calcutta, with three 540-hp Bristol Jupiter radial engines, a 93-foot span and an overall length of 66 feet, flew for the first time on 14th February, 1928. Five of these aircraft were built, the first being named the *City of Alexandria* and the second *City of Athens*, and these aircraft, with the three which followed, were intended for the Mediterranean sections of a UK–India route, operating to Karachi, in what is now Pakistan, and Calcutta. One of the aircraft was lost in an accident in 1929, but the other four remained in service until replaced by Short

Empire flying boats during the late 1930s. During their service life, Armstrong-Siddeley Tiger engines were substituted for the Jupiters, while a four-engined version of the Calcutta was the Short S.17 Kent, of which three were built for Imperial during 1931 and 1932. Even at this early stage, some movement was being made towards improving the performance and the safety margin of civil aircraft, for while the Singapore needed its two engines, the Calcutta could continue take-off on two engines if the third engine failed. Another Calcutta feature was a Fletner servo-rudder, the beginning of powered controls.

The British Government was obviously well satisfied with Imperial's performance, for the 1927–8 profits were up to £72,500, and instead of decreasing in accordance with the original plan, the subsidy was increased to £335,000. The airline had carried 11,395 passengers during its first year and 11,703 in the second year, but this then rose sharply to 17,083 in the third year and 26,479 in the fourth year. The Government also committed £415,000 towards the cost of improvements, which in many cases meant actually introducing air navigation facilities within the British Isles.

Ministers also appeared anxious to be seen to be associated with the growth of air services. Croydon Aerodrome had been rebuilt in 1928, and it was from the rebuilt aerodrome that the air minister, Sir Samuel Hoare, departed on 30th March, 1929, on the inaugural scheduled air mail and passenger flight to India. The air service was somewhat circuitous, even allowing for the limitations imposed by the need to refuel and by the poor reliability of the early aeroplane, because of arguments over landing rights and reciprocal air services. The minister and the other passengers left Croydon aboard an Armstrong-

A four-engined development of the successful Short Calcutta flying boat was the Short Kent for Imperial Airways.

Whitworth Argosy landplane, flying for five hours to reach Basle in Switzerland, before catching a train to Genoa in Italy, where they transferred to a Short Calcutta flying boat for the Mediterranean stage. This flew them via Rome, Athens and Tobruk to Alexandria, reaching the Egyptian port on 3rd April. There the minister switched to another aircraft for a tour of part of the Cairo–Cape route, while the rest of the passengers continued their journey aboard a de Havilland Hercules landplane, flying to Karachi, which they reached on 6th April. The route was extended to Calcutta later that year.

Not that everything could be expected to go smoothly. The one Calcutta to be lost, the *City of Rome*, on 26th October, 1929, was again on the Mediterranean section of the London–India route. The aircraft ran into a 70-mph gale on take-off from Naples and was forced to land in the sea, sending an SOS which was picked up by an Italian tug. The aircraft's captain kept the engines running to retain some control and also to ease the pressure on the tow line after this had been fixed. The crew of the tug heard the engines running for some fifteen minutes after the tow started, then lost contact in appalling weather and darkness. The aircraft was later recovered, but although the hull had remained watertight, the wind and waves had smashed the wings, causing it to capsize with the loss of the crew and the four passengers. A line breaking is a hazard in any towing operation, and the windage of the aircraft's wings could well have created problems in spite of the captain having had the presence of mind to keep the engines running. Even in calm seas flying boats were difficult to tow, drawing little water and in several instances riding high on the waves and colliding with the towing vessel.

Trippe, meanwhile, had seen his Long Island Airways collapse. Nevertheless, Trippe, his friend Hamilton and several like-minded associates, founded a landplane operation called Colonial Airways, operating passenger services but aided by the award of Civil Air Mail Route No. 1, New York to Boston, by the United States Government. Another company founded by Trippe at this time was the grandly named Aviation Company of the Americas, but while many of the pioneering companies of the period were to grow into the major American domestic airlines of today, Trippe's interests lay elsewhere. Trippe and the Aviation Company of the Americas were not to have a long association, for he next acquired two competing Florida-based airlines, Florida Airways and Pan American Airways, both of which were trying to gain a contract to carry air mail over the ninety miles between Key West and Havana. Trippe merged the two airlines, using the Pan American Airways name for the combined company, and he resolved any doubts over whether or not he would receive the United States Government air mail contract by negotiating with the Cuban authorities for the sole landing rights in that country. However, while being awarded Foreign Air Mail Route No. 1, Trippe was without a suitable aircraft, and so operations began using a leased West Indies Aerial Express Fairchild FC–2 floatplane, charging the United States Post Office $175.00 for a return mail flight between Key West and Havana. On the first flight, which took an hour, the aircraft carried seven bags of mail and arrived an hour earlier than the Cubans expected, so that the pilot spent some four hours attempting to convince the Cuban Customs officials that he was in fact on an official air mail flight!

Already Trippe was looking ahead, and far beyond the Caribbean. During the late 1920s, he moved through Latin America, securing landing rights whenever possible, and preferably on an exclusive basis. This meant that the US Post Office had to award air mail rights to the airline, as the only company which could provide not simply the lowest cost, but also demonstrate the ability to introduce a service with the least delay. There was only one serious competitor to the enterprising Pan American, the well-established United States shipping line, W. R. Grace & Company, which had considerable strength on the Pacific coast of South America, and was planning to develop air services of its own, using its knowledge of the area and its reputation. However, Trippe side-stepped this problem by forming a joint venture for the South American operations, Pan American Grace Airways (PANAGRA). Many years later, Pan American, which had eventually acquired complete control of PANAGRA during the 1960s, sold the airline to Braniff, which had also developed air services to South America, but this was well after the end of the flying boat era.

Ambitious plans could only be fulfilled if the right aircraft was available, and, searching for a suitable aircraft, the amphibian appealed to Trippe for his Caribbean services, largely because of the need to be able to use airfields in Cuba for emergency landings while over-flying the island, even though Pan American was normally as tied to water landing sites as was Britain's Imperial Airways or the French airlines. There was certainly no doubt that marine aircraft were suitable for Pan American Airways, or PAA as it was more usually known (rather than the modern Pan Am), for in spite of earlier experience with the landplane when working with the Dutch airline KLM, Pan American's chief engineer, Andre Prieste, contended that 'a seaplane carries its own airport on its bottom'.

However, a solution was at hand, coming from the Russian *émigré*

The S–34 flying boat, one of Igor Sikorsky's less successful designs.

Igor Sikorsky. Sikorsky had arrived in the United States in 1919 from France, where he had fled in 1917 at the height of the Russian Revolution. With the help of other Russian *émigrés*, Sikorsky had founded his own company in 1923, building a number of floatplanes, including a five-seat machine for a company exploring for oil in South America. Less successful was his S–34 of 1926, a twin-engined, twin-boom, high-wing monoplane with open cockpit, which was lost as a result of engine failure during a test flight, though fortunately without casualties. A successor was the S–36, an amphibian with two engines and an enclosed cabin for eight passengers, and one of the five S–36s built by Sikorsky was loaned to Pan American for trials. Although the aircraft was under-powered, Pan American was impressed with the reliability of the S–36, and used this machine to pioneer several South American and Caribbean routes, even though the hull design was still primitive and unsticking during take-off could be a problem.

During 1928, work started on a ten passenger amphibian powered by two Pratt and Whitney Wasp radial engines of 400hp each, the S–38. This aircraft retained the wing of the S–36 and a strengthened hull, and the end result was referred to rather unflatteringly as either the 'ugly duckling' or the 'flying tadpole'. A sesquiplane configuration helped to provide additional lift, while for simplicity, the undercarriage folded into the hull rather than retracting completely. The hull itself was divided into watertight compartments for greater safety. The technology of the day, the 'state of the art', meant that, excellent aircraft though it was, the S–38 had its share of shortcomings. Handling on the water was poor, while a vertical

A long relationship between Sikorsky and Pan American Airways started with the S–38 amphibian.

windscreen meant that water collected on the screen, during take-off. The structure also suffered from the effects of electrolysis, with brass nuts fastened into duralumin resulting in early and rapid corrosion, a process which was hastened in an aircraft operating to and from water. Strangely, Sikorsky described the aircraft as an 'amphibion'.

The S–38 led to a number of developments, including the single-engined S–39 and twin-engined S–39A, while the S–39B reverted to one engine, but all of these aircraft were 'mini' S–38s, with one of the operators being the airline Varney Speed Lines. Pan American bought seven examples of an improved S–38, the S–41, which also entered service with SCADTA and the United States Navy, in very small numbers. However, not everyone liked the S–41, which was criticized for having a 'step' too far forward in the hull, creating additional spray and handling problems.

During this period of rapid growth at Sikorsky and great interest in Sikorsky designs by potential customers, United Aircraft acquired the company, although it continued, and continues to this day, to use the name of its founder.

Pan American had not been the first airline to use the S–38, for that distinction was held by the amphibian and landplane operator the New York, Rio and Buenos Aires Line, NYRBA, or 'near beer' to humorists, one of the few significant competitors to Pan American. This company operated the S–38 alongside landplanes and the Consolidated Commodore flying boat, a civil development of BuAer's XPY–1. In a similar manner to the operations of Imperial Airways, NYRBA passengers changed from flying boat to landplane or amphibian for different stages of their journey, usually with landplanes providing connecting services across South America to NYRBA's trunk route up the Atlantic coastline, or between New York and Florida. However, before long, Trippe acquired this last main competitor, but not before rivalry between the two companies had reached a difficult and sometimes ugly situation at station level, not unlike the opposition between workers of rival railway companies during the nineteenth century. By 1931, Trippe had also taken an interest in SCADTA for $1 million, enabling his airline to operate between Miami and Barranquilla, while also benefiting from the connecting traffic within Colombia.

Having acquired his chief engineer from KLM, Trippe obtained his technical adviser from TAT, a predecessor to Trans World Airlines, and this was none other than the transatlantic solo flight pioneer, Charles Lindbergh, who joined Pan American in 1929. It was Lindbergh who flew most of the route-proving flights for Pan

When Pan American acquired the New York, Rio and Buenos Aires Line, it also accepted that airline's Consolidated Commodore flying boats.

American, for example, taking an S–38 on 4th February, 1929, from Miami to Havana, Belize, Managua in Nicaragua, Puntarenas in Costa Rica, and Cristobal in the Canal Zone, on what was to become Foreign Air Mail Route No. 5, operating over a distance of 2,300 route miles. Lindbergh was later to marry the daughter of the American ambassador to Mexico. A further survey flight by Lindbergh and Pan American's chief pilot, Basil Rowe, in mid-April, 1930, led to an inaugural 'mail only' flight on 26th April, 1930, operating from Miami on the Florida mainland rather than from Key West, to Havana and then Nicaragua and Cristobal, becoming the world's longest regular over-water air service, although, of course, it was anything but non-stop – out of the 1,294 miles over the Caribbean, just 630 miles was actually non-stop. However, this route was more direct than the original, cutting 200 miles off it, and even allowing for overnight stops and changing aircraft. It gave an eight-day New York–Buenos Aires flight compared to twenty days by sea. The first flight had not been without its problems, one of which was that it was raining so hard on arrival at Cristobal that Lindbergh had to stand on the rudder pedals, opening the cabin roof and peering over the top as the rain built up on the flat windshield to such an extent that neither Lindbergh nor Rowe could see ahead – fortunately, Lindbergh was tall enough to do this!

Its expansion and acquisitions also developed Pan American into a mixed aircraft operator, with its adoption of the NYRBA policy of operating flying boats on the Atlantic coast of South America and within the Caribbean, while the connecting flights across the Andes and within the United States were operated by landplanes, usually Fokker F.VII/3M tri-motors. By 1931, Pan American operated seventy-eight aircraft of all types on a 22,000-mile route network throughout Latin America.

There were hazards in operating amphibians rather than landplanes or flying boats, since pilots sometimes forgot to lower the landing gear, and at this time the landplanes had fixed, non-retractable undercarriages anyway. Rowe recalled, in his book *Under My Wings*, seeing a pilot landing an amphibian with the undercarriage up, and the aircraft overturning after hitting a rock on the runway. While upside down, the pilot lowered the landing gear so that he could claim that he had struck an obstacle during a correct 'wheels down' landing! Ironically, the story has it that the pilot in question was one who had previously been especially strident in his criticisms of those colleagues who belly flopped, and he was an accomplished landplane pilot himself.

Not all flights were routine. Apart from the eccentricities of the pilots, other factors sometimes took a hand. Twice during 1931, Pan American's S–38s found themselves on rescue and relief missions. On 31st March, a report was received from Managua of an earthquake, followed by complete radio silence. A Pan American S–38 on a routine Miami–San Salvador flight was ordered to set down its four passengers at San Lorenzo and then proceed to Managua to place itself at the disposal of the Nicaraguan Government. The aircraft found the city in flames, and its main task was to provide a radio link with the outside world while a rescue mission was organized. It later helped further by evacuating almost a hundred casualties in a series of flights. Relief personnel and medical aid were also flown into Belize, in what was then known as British Honduras, during September, 1931, aboard an S–38. A year later, a Pan American S–38 was commandeered to evacuate a Cuban dictator who had been overthrown, taking off just as an angry crowd rushed into the airfield. Another S–38 was stolen by a band of thieves led by a German living in Brazil, who hoped to sell the aircraft to a revolutionary movement, but unfortunately, one of his accomplices, a mechanic who had claimed to be a pilot, had lied, and no one else aboard could handle the aircraft. So, having taken off, it crashed after some five minutes in the air, with the loss of all aboard.

S–38s enjoyed an interesting career elsewhere, and in some cases their service life was a long one for aircraft of the day. Another S–38 operator was Inter-Island Airways, a predecessor of Hawaiian Airlines, and one of their aircraft, *Kauai*, was the only civil transport to escape damage when the Japanese attacked Pearl Harbor in 1941, afterwards flying meat into many of the islands during the war years. Other operators included Colonial and Western Airways, Western Air Express, and even companies such as Hughes Products and Standard Oil.

The S–38 and S–39 were also pioneering aircraft in the truest sense of the word, playing a part in aerial exploration. One private owner, Robert McCormick, used an aircraft named *Untin Bowler* for a transatlantic flight, but the aircraft was crushed by icebergs before leaving Newfoundland. McCormick replaced it with another S–38, named *Arf Pint*, but no more long-range flights were attempted. Another attempt at a transatlantic flight was by George Hutchinson in an S–38, *City of Richmond*, who planned to fly across the Atlantic with his wife and two young daughters. The public outcry that followed the announcement of Hutchinson's plans caused the Danish Government to ban the flight from Greenland and Denmark, the former being a vital staging point. However, undaunted, the Hutchinson family left Floyd Bennett Field on 23rd August, 1932, flying five hundred miles to St Johns, where the aircraft lost its wheels so that there would be weight to spare for extra fuel. They were then weather-bound for six days before the next stage, but on 30th August, they flew to Labrador, and on 2nd September, despite the Danish ban, flew six hundred miles to Greenland, where Hutchinson was fined KR 1,000, far less than the maximum fine of KR 5,000 for each of the aircraft's eight occupants, crew and passengers, that the Danish authorities could have levied. Little more than a week later, flying along the coast of Greenland, fog and ice caused a forced landing close to a small island, with the aircraft being wrecked against the rocks of the island after its occupants had scrambled safely ashore. Fortunately, they were soon picked up by a British trawler. The loss of their aircraft was unfortunate, but obviously Hutchinson possessed the ability to make a transatlantic flight, and could well have done so successfully had he started earlier in the summer.

Both an S–38, named *Osa's Ark*, and an S–39, named *Spirit of Africa*, respectively painted in zebra stripes and giraffe spots, were purchased by an American explorer, Martin Johnson, and his wife, Osa, for a trip to Africa. Both aircraft were shipped to Cape Town in 1933, and were subsequently used for aerial photography throughout East Africa. The S–39 was later renamed *Spirit of Africa and Borneo*, when used for a similar expedition to Borneo.

A small Sikorsky amphibian used by private owners as well as by airlines was the S–39, a scaled-down version of the twin-engined S–38.

One of the earliest pioneers of commercial transatlantic flight, although not in the strictest sense of the word, was the new German airline Deutsche Luft Hansa (DLH), founded in 1925, which commenced operations within Germany on 6th April, 1926, soon afterwards extending its services to a number of European capitals. This was the direct ancestor of the present German national airline, Lufthansa.

Even the aims of an ambitious airline such as DLH could not measure up to those of the German designer Professor Claude Dornier who, for all his practical work on such a successful and useful aeroplane as the Wal, was nevertheless planning something far larger, a large flying boat capable of crossing the vast ocean distances. The aircraft was his Dornier Do.X – the 'X' standing for experiment rather than for ten. A large monoplane, the Do.X used no less than twelve 525-hp Siemens-built Bristol Jupiter air-cooled engines in tandem pairs in six nacelles, each with a pusher and a tractor propeller, mounted above the mainplane. The aircraft used seawings, or sponsons, rather than wingtip floats. Designed to carry some seventy passengers on three decks, with lounge, bar, writing and smoking rooms, and with sleeping accommodation, the aircraft was built at Dornier's Swiss factory on Lake Constance. After a successful first flight on 25th July, 1929, the aircraft made a one-hour flight taking off from the Bodensee on 21st October, with 169 persons – 150 passengers, ten crew and nine stowaways! Problems with the air-cooled Jupiters driving the pusher propellers in the rear of the engine nacelles encouraged Dornier to switch to twelve liquid-cooled Curtiss Conqueror engines, with greater success. One feature of the aircraft, incidentally, was the provision of enclosed tubular steel struts leading from the mainplane to each engine, through which an engineer could pass to make an inflight inspection of the engines – although the Do.X was not the last flying boat to offer this facility, which now seems odd given the engine reliability of modern aircraft. Though the aircraft was mainly constructed from metal, the wings were still fabric covered.

Although under-powered and not suited to transatlantic commercial operation, with a maximum speed of 135 mph and a practical range of a little over a thousand miles, the Do.X was still an impressive aircraft, and one suspects sometimes that the knocking criticisms one hears of it come from the 'not invented here' brigade. Certainly, the Germans were anxious to show their masterpiece off to the world. On 2nd

The twelve-engined Dornier Do.X was many years ahead of its time, at least in terms of passenger capacity if not in technology.

An idea of the size and luxury of the Dornier Do.X.

In spite of its size, the controls of the Dornier, Do.X were still very basic.

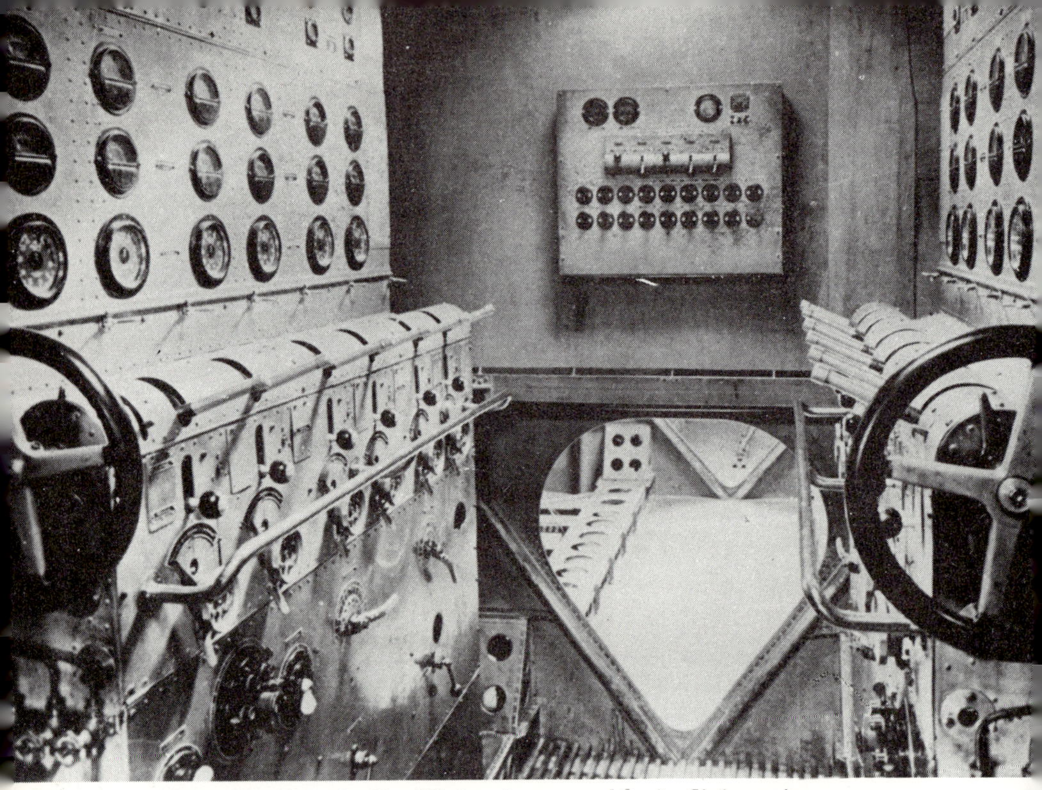

One of the Dornier Do.X's 'engine rooms' for in-flight maintenance, although this did not prevent the pusher engines overheating.

November, 1930, the huge flying boat left on a tour of Europe and the Americas, visiting Amsterdam, Calshot (near Southampton), Bordeaux, Lisbon, the Cape Verde Islands, Fernando de Novanha, Natal and Rio de Janeiro in Brazil, Antigua and Miami, before arriving in New York on 27th August, 1931. The flight was not an unqualified success, the aircraft being delayed on the way by fire damage at Lisbon and hull damage incurred while taking off in the Canaries. A further incident occurred in Brazil, when the captain was relieved of his command and returned to Germany after objecting to making goodwill exhibition flights or, as he is alleged to have put it, 'joy rides', but perhaps his objections have lost something in the translation! The aircraft wintered in an American dry-dock, but left on the first leg of its flight back to Germany on 19th May, 1932, flying via Newfoundland and the Azores to Lisbon, although due to a shortage of fuel the aircraft landed in the sea six miles west of Lisbon and completed the journey by taxiing into the River Tagus. DLH operated the aircraft briefly and, one imagines, unprofitably before passing it to the DVL, Deutsche Versuchsentace für Luftfahrt, research institution, but it was finally destroyed during an Allied air raid on its Berlin museum home during World War II. Nevertheless, two Do.X flying boats were built for Italy, using 550-hp Fiat water-cooled engines, and these aircraft, *Umberto Maddelene* and *Alessandro Guidoni*, were to be introduced on the Mediterranean operations of the S.A. Navigazione Aerea, though were in fact never used as such. The Regia Aeronautica operated them experimentally before they were scrapped. The design was obsolete, being essentially an enlarged Wal, but it was also an instance of an aircraft being too big too soon. Fewer, larger engines, had they been available, would have done much for the economics of the aircraft, but at the time it was too much to expect sixty or seventy people to board a long-distance flight, the market just wasn't big enough.

Dornier could afford a flop. His factories were busy building the Do.J Wal, the uprated Do.15 Wal and the Do.R Super Wal, for a number of airlines, with CMASA in Italy, Aviolanda in the Netherlands and Kawasaki in Japan, building civil versions as well as military ones. Many of the airlines opted for the larger 'Ten Ton Wal', including DLH and Aero Lloyd in Germany, Aero Expresso and SANA in Italy, Condor and Varig in Brazil, SCADTA in Colombia and Nihon Koko in Japan, carrying eight to ten passengers per flight. The standard Wal was also used by the Condor Syndicate, which was backed by DLH for operations throughout South America. Relatively few Super Wals were built, although DLH also used this nineteen-passenger aircraft, powered for its test-flight by two 650-hp Rolls-Royce Condor engines and subsequently more usually powered by four engines in two nacelles.

The Wal and Super Wal featured in DLH transatlantic air mail operations during the 1930s. These were not transatlantic flights, but were enterprising and, within their limitations, successful nevertheless.

On the North Atlantic, DLH commenced operations with catapult-launched Heinkel He.12 and He.58 biplanes, which were replaced in 1932 by Junkers Ju.46 monoplanes catapulted from the liners *Bremen* and *Europe* when some 750 miles from their destination, helping to accelerate the mail service.

Even more ambitious, using two catapult-equipped cargo ships, the *Schwabenland* and the *Westfalen*, the Wal operated from 1934 on a mail service from Bathurst, in the British colony of the Gambia, to whence mail had been flown from Germany to Brazil. The aircraft did not fly for the whole distance, departing aboard one of the two cargo ships and not being catapulted off the ship to fly with a small quantity of mail to Rio de Janeiro until the vessel was some distance out to sea. However,

As well as operating on the South Atlantic, Luft Hansa cut mail times on the North Atlantic by flying the last 750 miles in Junkers Ju.46 seaplanes, launched from liners at sea.

this system did cut the transit time for mail travelling from Germany to Brazil to as little as three days. The other ship would operate the return service in a similar manner, sailing from Rio and then launching the aircraft, and mail, on its flight to Bathurst. Throughout several years of operation, some three hundred flights were completed without accident. In 1936, Air France joined DLH on the service to eliminate costly competition. The Wal and Super Wal did not have the service to themselves, with the stylish Blohm und Voss Ha–139, four Junkers Jumo diesel-engined monoplane seaplanes, joining the service in 1938, along with the Dornier Do.26 *Seeadler*. The Dornier also used four Jumo engines, although in two nacelles with tractor and pusher propellers, the latter having its engine 10 degrees off the horizontal to keep the propellers clear of the water during take-off. The Do.26 was one of the few Dornier designs not to have sponsons, having retractable wingtip floats instead. It was able to fly at up to 200 mph and over ranges of up to 5,000 miles.

The French had also worked independently on ship and air mail services which, while not providing through air mail flights, did, nevertheless, accelerate the mails. In 1928, a predecessor of Air France, the Compagnie Generale Aéropostale, initiated an accelerated transatlantic mail service, operating from France to Dakar in West Africa by air and transferring the mail to a destroyer for the

A Dornier Super Wal or 'Ten Ton Wal' is catapulted on a South Atlantic mail flight.

The Super Wal took some handling.

The size of the Super Wal can be judged from this shot of the launching platform on the *Westfalen*.

The elegant lines of *Seeadler*, one of Luft Hansa's Dornier Do.26s. The pusher propellers can hardly be seen in this photograph.

A nose-on view of the Do.26, showing the retractable floats.

transatlantic crossing to Natal in Brazil. Delivery of the mail throughout South America was usually by air.

However, the French also attempted a direct air mail crossing of the Atlantic. On 11th May, 1930, Jean Mermoz, a French airman, took off from Dakar in a single-engined Latécoère 28 high-wing monoplane which had been converted from a landplane configuration to a seaplane, and with two companions, Mermoz flew this aircraft, named *Comte de la Vaulx*, on a 19 hours, 35 minutes, non-stop flight to Natal with a small consignment of mail. The flight established a non-stop distance record for hydro-aeroplanes. Much later, a return flight was attempted from Rio to Dakar, but while some five hundred miles off the African coast, an oil leak forced the aircraft down into the sea, and while the crew and the mail were saved by a rescue ship, the aircraft itself lost a float and sank in a heavy swell. Wisely, Aéropostale decided that the time was not right for a transatlantic air mail service. Indeed, the company went bankrupt in 1931 and was acquired for merger into the new Air France in 1933. However, Mermoz made a successful landplane crossing of the Atlantic before this.

The Atlantic, with its lack of good staging points, was a greater challenge than the Pacific in some ways, even though the overall distances between major cities tended to be less than those over the Pacific. In the North Atlantic, separating the largest cities of Europe and the Americas, uncertain weather conditions were another factor threatening the establishment of regular, and reliable, air services.

Air France had hoped to use its Breguet Br.530 Saigons, developed from the Calcutta, on a transatlantic service, using an uprated version to be called the Dakar. However, these plans were abandoned before the first aircraft was built. Meanwhile, the first Br.530 Saigon, called *Algérie*, flew in May, 1934, and by January, 1935, *Algérie* and a second aircraft, *Tunisie*, were operating from Marseilles to the Mediterranean points of the Air France network, carrying twenty passengers in three cabins; two or three travelling de luxe aft, six in first class amidships, and eleven in second class in a forward cabin. Using three 785-hp Hispano-Suiza engines, the Saigon could fly at 140 mph.

In Italy, naturally enough, trans-Mediterranean air services also attracted airline attention, in spite of the more far-flung activities of Balbo. As early as August, 1926, Aero Expresso Italiana operated a regular Brindisi–Constantinople (now Istanbul) service using the Savoia–Marchetti S.M.55C, the commercial version of the aircraft used by Balbo for his exploits. The S.M.55C carried five passengers in each of the two hulls. The civil version was also improved as the years passed, and the S.M.55P of 1928, with twin uprated 500-hp Isotta–

Fraschini Asso engines and hulls stretched to take six passengers in each, proved popular with the Italian airlines, with fifteen being delivered to SAM, the Società Aerea Mediterranea, while no less than twenty-three of the next version, the S.M.66, with three 750-hp Fiat engines and a maximum speed of 160 mph, were sold to SAM, Aero Expresso Italiano and Ala Littoria, with these aircraft taking nine passengers in each hull.

The general trend in flying boat development, as in most things, was upward, with successive aircraft being larger and faster, although throughout the history of the flying boat the increase in size was always more marked than any increase in speed. However, larger or faster, there were still many problems to resolve before the flying boat could reach its full potential. In developing his next aircraft for Pan American, Igor Sikorsky devoted considerable effort to evolve a safer and more efficient hull form. While Shorts were building testing tanks in which scale models of flying boat hulls could be evaluated, Sikorsky, also building scale models, towed his behind a speedboat on the Housatonic River, making some 2,000 runs at speeds of up to 30 knots with several different hull forms, which were generally some six feet in length. At the same time, rather than undertake a complete redesign with all the risks which that involved, Sikorsky preferred to use much of the S–38 design in the new S–40, disappointing Pan American's Lindbergh, who expected to see much more streamlining, but doubtless reassuring many others. Indeed, Lindbergh referred to the S–40 as the 'flying forest', a reference to the large number of struts in this large sesquiplane with four 575-hp Pratt and Whitney Hornet radial engines mounted on the struts between the wings. A twin-boom tailplane extended from the upper wing trailing edge. While Sikorsky had his way on most things, Lindbergh, who always pushed the pilot's point of view, insisted on moving the flight deck forward, clear of the spray, and this was done at an early stage in the development of the aircraft, so that all S–40s offered the pilots excellent visibility during take-off and landing.

One of the main problems with the S–40 concerned neither the hull nor the position of the pilots' cabin but resulted from Pan American's insistence that the aircraft should be an amphibian rather than a flying boat in case of emergency landings while flying over land – a reversal of the usual fear of ditching in the sea which always favoured the flying boat configuration. The S–40 was the largest aircraft built in the United States at that time and Sikorsky had great difficulty in finding springs large enough for the main undercarriage of the 114-foot wingspan and 77-foot-long aircraft. The problem was solved, however, by using railway carriage springs for the 17-ton aircraft. In the event, the reliability of the aircraft and its engines was such that, while the first three aircraft were built as amphibians, the remainder were completed as pure flying boats.

First flown in spring, 1931, the S–40 could accommodate the then staggering total of up to forty passengers and one ton of cargo, mail and baggage, over distances of up to 950 miles. The first S–40 was officially named *American Clipper* by the wife of the President of the United States, Mrs Herbert Hoover, who, because of prohibition, had to use a bottle of water from the Caribbean for the official naming ceremony rather than the more usual champagne. From this time on, all Pan American aircraft became known as 'Clippers', which the airline registered as a trade name, while the pilots became 'captain' and the co-pilots became 'first officer', perpetuating a merchant navy image. However, unlike Instone, the Pan American aircrew uniforms looked more like those of well-to-do yachtsmen.

On most of its services, Pan American reduced the seating capacity of the S–40 to thirty-two passengers, and it was in this form that the amphibian made its first revenue-earning flight from Miami to Kingston in Jamaica, and on to the Panama Canal Zone. The aircraft

The Sikorsky S–40 was known as the 'Flying Forest' – one can see why!

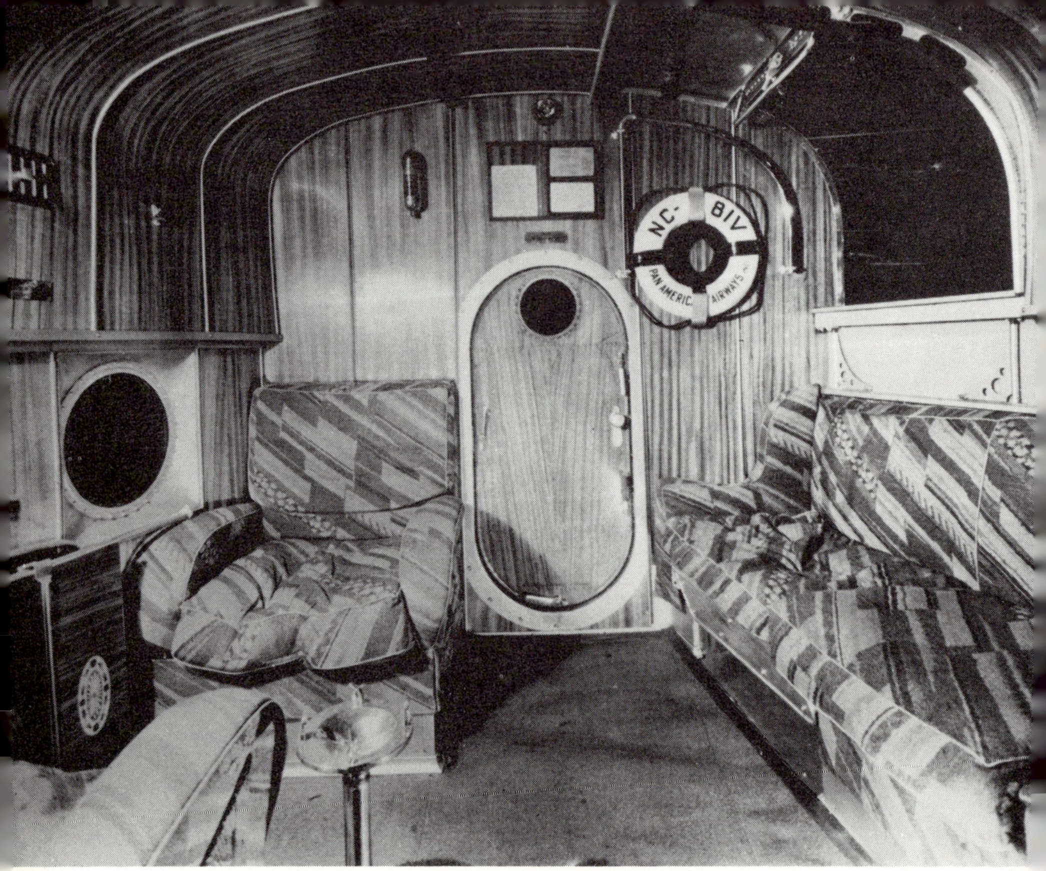

Interior comfort on the S–40, used on Pan American's Latin American routes.

could not be flown at night due to the lack of night flying aids, and in spite of all the trials, it needed careful handling, especially during landing, because of a marked tendency to swerve to port. However, the aircraft flew successfully, and three Pan American S–40s 'joined' the United States Navy during World War II to become training aircraft for naval flying boat pilots.

Still higher performance was demanded from Sikorsky for the next aircraft for Pan American's services, setting a specification that demanded the ability to fly for 2,500 miles against a 30-knot headwind and a higher speed than any competing flying boat. This meant that the aircraft would need to have a still higher wing-loading, leading to the S–42, with 118-foot wingspan, although slightly shorter than the S–40. The S–42 differed significantly in appearance from its predecessor, with a conventional fuselage with a triple fin tailplane and a monoplane wing above the fuselage, which featured a centre flap for take-off and landing in addition to the normal outer wing flaps. The four 750-hp Pratt and Whitney Hornet radial engines were faired into the wing, while the aircraft could fly at 190 mph, with a normal range of 1,200 miles and up to forty passengers, although, again, thirty-two tended to be the more usual capacity. The manufacturer offered a fourteen-passenger sleeper version for overnight flights. During development of the aircraft, the 750-hp engines had replaced lower-rated 700-hp versions of the same engine, and the step in the hull was moved aft to improve handling during take-off and landing. Ice in the Housatonic River delayed tests of the prototype from Christmas, 1933, to March, 1934, but it soon became apparent that while the performance might not be quite that which Pan American had specified, it did nevertheless meet their immediate needs, and the aircraft generally handled well. The real point had to be that it was ready for service far earlier than the rival Martin M–130, so that Sikorsky was able to proceed with the construction of ten aircraft for Pan American, including three standard S–42s, four S–42As, and three S–42Bs, the last having additional fuel tanks for transatlantic or trans-Pacific operation. The first aircraft was delivered on 5th June, 1934, and on 15th August, departed on its first revenue-earning flight to Buenos Aires, being named *Brazilian Clipper* by the wife of the Brazilian president while refuelling at Rio de Janeiro.

A rather more sleek design was the Sikorsky S–42, most of which were completed as flying boats rather than as amphibians.

The relatively modern interior of the S–42.

The S–42 managed to create a series of record-breaking flights, setting ten payload-to-altitude records for flying boats. The remaining seven records available at the time were held by a French aircraft. As well as Sikorsky's chief test pilot, Boris Sergievsky, and Pan American's Charles Lindbergh, the other pilot on these first flights was Pan American's chief pilot, Captain Edwin Musick.

As aircraft became larger, a fresh need was created for smaller aircraft for the less busy and shorter routes, leading to a scaled down version of the S–42, the twin-engined S–43 amphibian, which also acted as an S–38 replacement, accommodating up to eighteen passengers and being introduced on Pan American's Caribbean network. The S–43 was also purchased by Inter-Island Airways, by PANAGRA on its feeder routes and by Panair do Brasil, whose domestic routes linked into the Pan American Airways international services. The USN and the USAAC also put the S–43 into service.

The small sister of the S–42 was the Sikorsky S–43 amphibian, for the less busy routes. This was a Pan American aircraft.

The Martin M–130 possessed the range necessary for transatlantic and trans-Pacific services, carrying a full payload. It had been the only competitor to Sikorsky, since other manufacturers approached by Pan American had refused to tender. Trippe ordered just three M–130s, and after the first of these had been named *China Clipper*, the name applied to all M–130s, even though Pan American's second and third aircraft were named *Hawaiian Clipper* and *Philippine Clipper*. Designed with a range of 3,200 miles, the Martin M–130 was a monoplane with sponsons, using four 830-hp Pratt and Whitney Twin Wasp radial engines, two of which had to be changed every three trips to ensure engine reliability on the long over-water distances – an exacting maintenance requirement which reflects the high number of flying hours of this aircraft by the standards of the day. The normal cruising speed was 150 mph. First flown in early autumn, 1935, the first aircraft was delivered to Pan American on 9th October, 1935.

Known collectively as the 'China Clippers', the Martin M–130 was designed for the long Pacific routes.

Trippe had been inspired by the idea of transatlantic flight as early as 1919. Of course, for many years the idea was an impossibility, but even so, by 1929, Trippe felt confident enough to press Imperial Airways and Aéropostale to consider joint transatlantic services. The idea was that three routes should be considered, a northern or Greenland route, a southern or Azores route, and a 'Great Circle Route' between these two, and that the northern route should be jointly operated with Imperial and the southern with Aéropostale. A fourth route was also considered, to Copenhagen, and this was to some extent a spur off the northern route for passengers flying to Scandinavia and Germany. During 1933, Lindbergh and his wife used a floatplane version of a Lockheed Sirius monoplane to survey the three main routes, having earlier used this same aircraft to survey routes along the coast of China and to Japan from Alaska. Problems with landing rights in the Azores, which were part of Portugal, and in

The spacious interior of the Martin M–130.

Newfoundland, which by this time had become part of Canada, delayed plans for a North Atlantic service.

Meanwhile, Air France and Imperial Airways were looking to the Far East and Africa for their development. Trippe too was considering the Far East, and discouraged on the North Atlantic, he decided to make the best of the situation and inaugurate a trans-Pacific route. In a sense there was logic in this, for Hawaii, although not then an American state, was important commercially and militarily, and so too were the Philippines, with which the United States had close ties.

Rather than simply linking the Philippines and Hawaii with the United States, Trippe took a more ambitious course, planning a trans-Pacific route to China. As before, in Latin America, he secured his landing rights before completing his plane for the service, acquiring the China National Aviation Corporation, which operated in opposition to the Luft Hansa-supported Eurasian Aviation

Corporation. He re-equipped the airline with a fleet of Douglas Dolphin amphibians to operate its network of services in China, a network which could feed traffic into a trans-Pacific service. A new operating base was formed at Alameda, close to San Francisco, for trans-Pacific flights, where the S–42s and M–130s could be serviced, and a ship, the SS *North Haven*, was chartered to carry construction teams to the Pacific islands of Midway, Wake and Guam – vital staging posts for the long trans-Pacific flight – to build bases and hotels for passengers.

The first survey flight of the new route was commanded by Captain Edwin Musick, starting on 16th April, 1935, and using an S–42 for the 18-hour 9-minute, 2,300-mile flight from San Francisco to Honolulu, and returning on 22nd and 23rd April, 1935, although flying eastwards, high winds meant that the flight took 21 hours. A further flight on 12th June reduced the flight time to Honolulu to 18 hours, and on 15th June, the aircraft took off for the 1,320-mile flight from Honolulu to Midway, landing there just 9¼ hours later. Further flights followed, including some from Midway to Wake and Guam, and on one flight the aircraft flew with the flight deck windows blacked out to simulate night-flying conditions.

Eventually, Pan American was awarded Foreign Air Mail Route No.

Pan American used the Douglas Dolphin amphibian on feeder services along the coast of China.

14 to the Philippines, and on 22nd November, 1935, the *China Clipper* took off on the first air mail service to the Philippines, carrying mail only, some 110,000 items weighing 1,837 lbs which were brought to the aircraft by an old 'Wild West' stage coach. Doubtless to the relief of construction workers building the Golden Gate Bridge, plans to fly over the bridge on take-off were abandoned and the aircraft took off under the bridge for its 21-hour flight to Hawaii. Out of seven crew aboard no less than five, including Captain Musick, were pilots. The flight arrived at Honolulu on 23rd November, and the following day, a Sunday, departed early for Midway, arriving in mid-afternoon. They took off again early on Monday morning for Wake, flying across the International Date Line so that they arrived at Wake the following day! On 27th November, they flew to Guam, arriving after lunch only to find that they were a day early since the station personnel had forgotten about the International Date Line! Eventually, they arrived in Manila on time on 29th November. The return flight, departing from Manila on 2nd December, meant flying through squalls on the edge of a typhoon between Guam and Wake. A day later they regained the day lost when flying west through the Date Line, eventually to arrive at San Francisco on 6th December, with 108,000 letters.

In fact, a few passengers had been carried on certain stages of the outward flight, relief personnel for the hotels at Guam and Wake, and some canaries to relieve the silence at Wake Island. The flight had taken 59 hours 48 minutes westbound and 63 hours 24 minutes eastbound. Passengers were first carried on the flight departing San Francisco on 21st October, 1936, with the *Hawaiian Clipper*. They paid $799 single for the through trip to Manila, or $1,438 return. The third flying boat, the *Philippine Clipper*, had the distinction of making the first through flight with passengers to Hong Kong, leaving San Francisco on 2nd November, but the usual practice was to use an S–42, *Hong Kong Clipper*, on this stage since it was more economical than flying the large M–130, with up to 52 seats, over the entire route. The China terminus had been switched to Hong Kong for commercial reasons, but only after local business interests had protested at the thought of the flight using a Chinese port, causing the British authorities in Hong Kong to grant Pan American the necessary landing rights.

Further route expansion in the Pacific was planned before the outbreak of World War II in 1939 caused these plans to be abandoned. On 17th March, 1937, the S–42 *Pan American Clipper II* flew from San Francisco and Honolulu to Kingman Reef, Pago Pago, and then to Auckland in New Zealand, arriving there on 29th March, at the same time as the Imperial Airways flying boat *Centaurus* – one of the Short Empire 'C' class flying boats on a survey flight from Australia for a joint Imperial Airways and Qantas UK–India–Australia–New Zealand route.

However, this steady development, even taken in stages, as indeed it was, could not be without incident and loss. On 11th January, 1938, Captain Musick and his crew were lost while on a survey flight in the S–42 *Samoan Clipper* off Pago Pago, while preparing for the Pan American service to New Zealand. Later that year, the M–130 *Hawaiian Clipper* was lost between Guam and Manila, with all six passengers and nine crew, and no trace was found of them or of their aircraft. Later, *Hong Kong Clipper* was destroyed by Japanese shelling while at Midway during the early stages of the war in the Pacific, and before this, an S–42 had crashed in the bay at Rio de Janeiro, with only two out of the sixteen persons aboard being rescued.

The same pressures for longer range aircraft also applied on the other side of the Atlantic, where Imperial Airways not only had to plan for its own objectives in flying to India and Australia, and eventually across the North Atlantic, but also had to be able to meet its obligations under the Government's Empire Air Mail Scheme. Conceived in 1934, the Empire Air Mail Scheme was a British Government plan to ensure that all letters between the countries of the British Empire could be sent by air, at the rate of 1½d (0.625p) per half ounce, or 2½ US cents at the exchange rate then current. With the quantity of mail likely to be generated by the scheme, Imperial Airways approached several British aircraft manufacturers to see whether they could provide an aircraft to meet the technical specification drawn up by the company's technical adviser, Major R. H. Mayo. Only Short Brothers responded, and Imperial took the then unusual step of ordering twenty-eight of the new Short S.23 flying boat off the drawing-board, at a total cost of £1.75 million, without a prototype aircraft. First flown on 4th July, 1936, the first of what was to be known as the Short Empire 'C' class was not a prototype but a standard production machine named *Canopus*, using four 910-hp Bristol Pegasus radial engines, with a maximum speed of 200 mph, a range of 760 miles fully laden, a wingspan of 114 feet and 88 feet long. A new feature introduced on this aircraft was the 'Gouge flaps', named after Shorts' chief designer, Arthur Gouge, and tested on a Short Scion, which increased wing area during take-off by 30 per cent. The initial order was soon increased to thirty-one aircraft, with six of these being intended for the Australian airline Qantas Empire Airways – known today simply as Qantas.

Cooee, one of the six Qantas Empire flying boats for the joint UK–Singapore–Australia service operated with Imperial Airways.

A gamble, but one which paid off, the Short S.23 Empire aircraft all received names beginning with 'C', which produced a weird variety of names. Qantas aircraft, for example, were named *Coolengatte, Cooee, Carpentaria, Coogee, Corio* and *Courang*, although some at least of these were obviously place names. Deliveries were made at the then incredible rate of two aircraft per month, with the aircraft entering service on 31st October, 1936. Such was the success of the aircraft and of the scheme that, although designed to carry twenty-four daytime or nineteen overnight passengers, the aircraft subsequently had to have their seating capacity reduced to seventeen to accommodate an extra 4,000 lbs of freight and air mail.

The Australian services had started with flights from Croydon to Brisbane, in Queensland, from 20th April, 1935, but with the arrival of the Empires, these aircraft flew the entire route, operating from Southampton. The Empires reduced the previous flight time to Brisbane from 12½ days to 9 days, and that compared with 30 days by a fast ocean liner. The Empire Air Mail Scheme was officially inaugurated on 29th June, 1937, and fully operational by 1938. There were some difficulties at first. Contrary to the rules of the scheme, the Australian Post Office surcharged air mail for Britain, making the cost 5d (2.08p) per half ounce – the same rate charged for Australian internal mail, in all fairness to the Australians, who did not see why overseas mail should be cheaper. Australian protests against the official rate were based on the fact that this would require a subsidy of £70,000 a year. In South Africa, rates were double the Empire rate. Such problems were not only inevitable, but in many ways have always defied solution, since air mail, and overseas surface mail as well for that matter, is charged by the country of origin at rates reflecting their costs and standards, while the country of destination makes deliveries free of any revenue from the mail, the cost of international transport having been paid by the country of origin.

Flight by the Empire 'boats' was possibly the most civilized form of air travel ever devised. Leisurely by today's standards, with few overnight passenger flights, passengers were able to relax in luxurious hotels, such as the Grande Bretagne in Athens or the Raffles in Singapore. Passengers were flown from Southampton, having travelled down from London by train the evening before, and then flew to Marseilles, Rome, Brindisi, either Athens or Crete, then to

Plenty of room aboard a Qantas Empire flying boat.

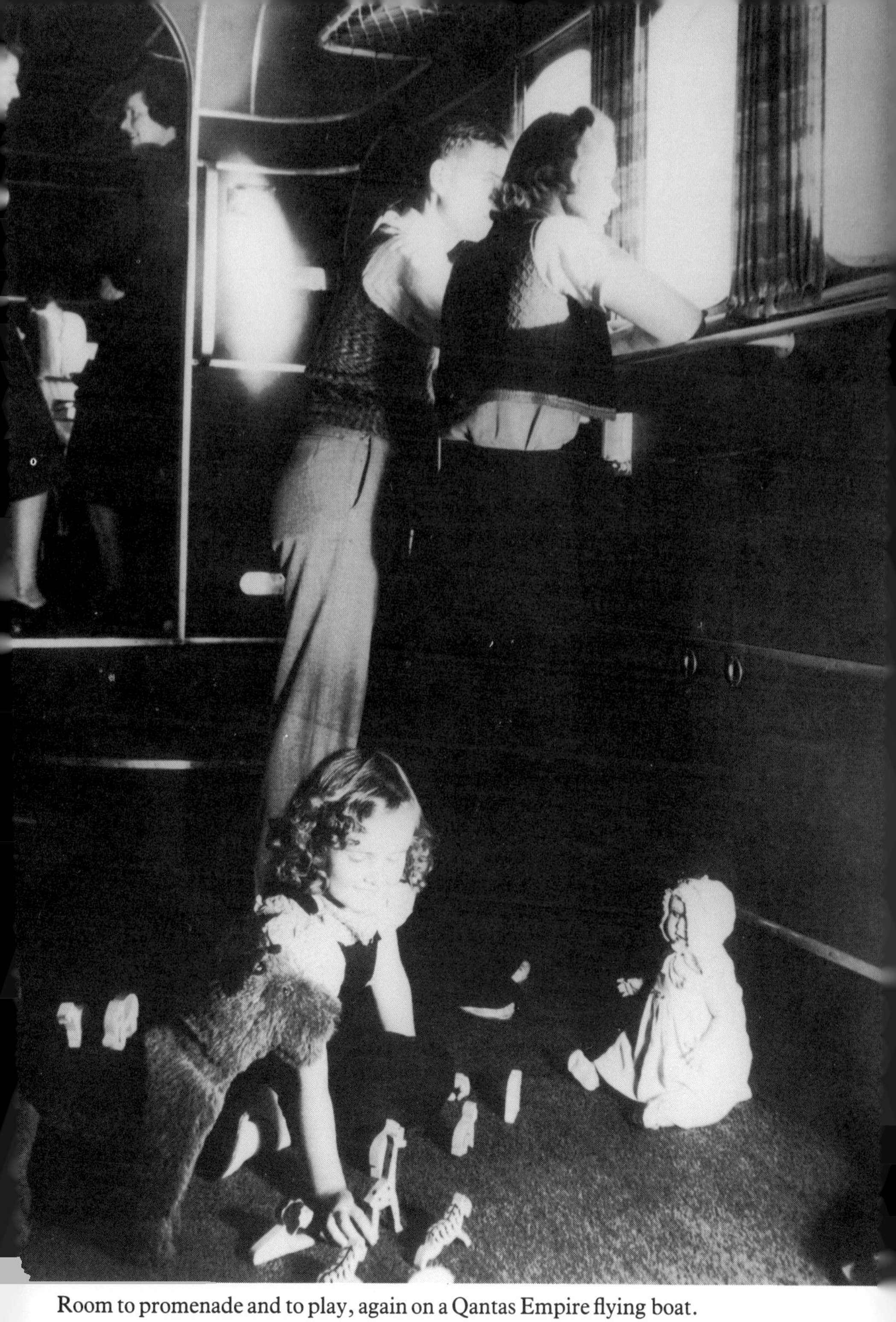

Room to promenade and to play, again on a Qantas Empire flying boat.

Alexandria, Tiberius, Habbaniyeh, Basra, Bahrain, Dubai, Gwadr, Karachi, Raj Samand, landing on a lake at Gwalior and then again at Allahabad, Calcutta, Ahyab, Rangoon, Bangkok, Penang, Singapore, Baatovic, Sourabaya, Rambang, Koupang, Darwin, Gooli Eylanalt, Karumba, Townsville, Gladstone, Brisbane and, finally, Sydney. At Bangkok, they would change to a de Havilland D.H.86 landplane for Hong Kong. Often, the one aircraft would not have flown the whole journey, with the change of aircraft usually at Alexandria in Egypt, but regardless of this, a Qantas crew would take over at Singapore – their white caps, dark navy blue shorts, tie and socks, and white uniform shirt with epaulets putting, as one traveller remarked, 'the Imperial Airways pilots' khaki in the shade'. Of course, as with passengers on ocean liners sailing to the Empire, one had to adopt a different lifestyle during the flight, with very early morning starts being the norm, but if one had stayed up too late the night before, there was time for a decent nap before a hot breakfast of bacon and eggs – one couldn't be expected to change one's lifestyle too much.

The London–Singapore single fare of £156 was reduced by 20 per cent in September, 1938, to £130, but the cost of the return fare was only £274 for the return London–Sydney flight, so perhaps it paid to stay the whole course! Such charging indicates that there must have been relatively few passengers joining and then alighting again at intermediate stops.

Unfortunately, such intensive operations and the limited navigational aids of the day, combined with the dangers of operating to and from water, meant that in the first two or three years of operation, almost a quarter of the Empire flying boats crashed, usually as a result of pilot error or adverse weather conditions. The rate of accidents was such that one aeronautical journal, *Flight*, now *Flight International*, saw fit to question Imperial Airways' pilot training.

One of the earliest accidents occurred to the tenth aircraft, *Capricornus*, which crashed in a snowstorm while flying near Mâcon in France on 24th March, 1937, her maiden flight on the Southampton–Alexandria service, with the loss of all four crew and with fatal injuries to the one passenger. The aircraft's crew had become disorientated in snow and hail, and was making for the emergency landing-place near Mâcon. Later that year, *Courtier* was lost at Athens on 1st October, with the loss of three lives, when the pilot failed to flatten out after being misled by the smooth surface of the water. *Cygnus* crashed on 5th December during take-off at Brindisi, with the death of a passenger and the steward, although most of those on board the aircraft escaped, including Air Marshal Sir John Salmond. The next accident was not until 27th November, 1938, when *Calpurnia*, on Christmas mail flight, crashed in a sandstorm into eight feet of water on the approach to the station at Lake Habbaniyeh, in Iraq and some fifty miles from Baghdad. Four crew were lost, and another two were seriously injured.

Sometimes lessons were learnt from the accidents. On 22nd January, 1939, *Cavalier* crashed while operating on the joint Pan American and Imperial Airways New York to Bermuda service, with the loss of three lives. Flying through a snowstorm some 150 miles off Bermuda, the aircraft landed heavily in the sea when the engines failed after icing of the carburettors, even though the air intakes had been heated. Two passengers and a steward were lost, and the survivors were in the sea in their life-jackets for ten hours. As a result of the accident, the Air Ministry required that all passengers must be strapped into their seats during take-off and landing, since one of the passengers who died had been fatally injured while attempting to stand up during landing. The emergency escape hatches were also modified to make escape to the top of the fuselage easier, and modifications were made to give the crew greater control over the engine air intake temperatures.

On 12th June, 1939, *Centurion* was lost and one passenger was fatally injured after it hit a mud bank during take-off on the Hoogly River, near Calcutta, but the other four passengers and four crew escaped injury, even though the aircraft sank. Before this, *Challenger*, on the South African service, had struck the shore at Mozambique, after hitting the water and bouncing to a height of ten feet, killing the flight clerk and the radio officer. This was in April, 1939, and was in fact the first serious accident on the South African service, to which the Empire flying boats had been introduced on 15th May, 1937. Not all of the accidents, of course, resulted in fatalities or even in the loss of the aircraft, but it was at this stage that *Flight* questioned the training of flying boat crews, and called upon the British Overseas Airways Corporation, which had been formed by the merger of Imperial Airways and British Airways as a recommendation of the Cadman Report of 1938, to improve standards. Initially, training had been given on flying boats of older classes, but as the Empire fleet had developed, the only training was alleged to be as the PT2, or co-pilot, on the Empires. On one occasion, an accident had occurred after the captain's request for 'flaps' was misunderstood, and the first officer gave full flaps rather than landing flaps, so that the aircraft ballooned out of control.

There were other types of accident as well. One of the strengthened flying boats for the North Atlantic services, *Connemara*, was lost in a refuelling accident, a fire, at Hythe, near Southampton, with the loss of one man, after the fire had started on a refuelling barge and spread to the aircraft. Rather luckier was *Corsair*, which made an emergency landing on the River Dangu in the Congo in March, 1939. The aircraft was repaired and took off again the following January, after a 2,000-yard road had been cut through tropical rain forest so that dams could be built in the river to provide a sufficiently large and deep enough stretch of water for a take-off!

A development of the S.23 was the heavyweight S.30, still part of the Empire 'C' class, using four 890-hp Bristol Perseus radial engines, with four of these being equipped for inflight refuelling. Three aircraft of another version, the S.33, were ordered, but only two were built before World War II broke out and the third aircraft was cancelled. However, three stretched S.26 'G' class flying boats, *Golden Hind*, *Golden Fleece* and *Golden Horn*, using four 1,380-hp Bristol Hercules engines, were built for non-stop North Atlantic operations, although these passed almost immediately into Royal Air Force service for VIP operations during World War II, with two being lost and the third, *Golden Hind*, destroyed during a storm in 1954.

Transatlantic flight had moved a stage closer in 1937 with agreement between Pan American and Imperial for a joint transatlantic service, providing both airlines could operate on such a service. During that summer, PAA's S–42 *Pan American Clipper III*, which had originally been named *Alaskan Clipper*, made an Atlantic survey flight, taking off from Port Washington and flying via Botswood, Newfoundland, to Southampton, while the S.23 *Caledonia*, of Imperial Airways, flew eastwards to Botswood. One early problem was that the prevailing winds, as the old sailing vessel masters had discovered, were westerly, making the westbound flight longer and more difficult for aircraft close to the limits of their endurance, than the eastbound flight. Initially, *Pan American Clipper III* took 12½ hours eastbound while *Caledonia* took 15 hours westbound, but many weeks of survey flying passed until the end of the trials for that year on 13th September, by which time *Caledonia* had improved upon her times.

There was no lack of initiative at Imperial Airways in extending the range of the flying boat. Undoubtedly, the most novel idea was the Short Mayo composite aircraft, the brainchild of Imperial's Major Mayo, which involved a variation of the Empire flying boats, the Short S.21, carrying a four-engined seaplane aloft to be released in mid-air to fly with a full load to its destination. The concept was to reduce the amount of fuel used by the seaplane during take-off by giving the aircraft a flying-start! The S.21 flying boat, named *Maia*, used four 910-hp Bristol Pegasus radials and had a broader-beamed hull than the S.23. Its passenger, the S.20 seaplane, named *Mercury*, the winged messenger, used four 340-hp Napier Rapier engines, and had a wingspan of 73 feet and was 51 feet long, with a 195 mph maximum speed and a range of some 4,000 miles with a payload of mail. During take-off, the composite aircraft used all eight engines, although the controls on *Mercury* were locked until the aircraft could be released. *Maia* first flew on 27th July, 1937, followed by *Mercury* on 5th September. The first flight of the composite aircraft was not until 20th January, 1938, with the first mid-air launch of the seaplane on 6th February, 1938. Non-commercial such a costly combination might have been, but it worked, and indeed worked very well.

On 21st July, 1938, the Mayo took off from Foynes in Ireland to launch the *Mercury* on a non-stop flight to Montreal which took 20⅓ hours, carrying 600 lbs of mail. On 6th October, 1938, an official 4,000-mile range notwithstanding, the Mayo took off from Dundee in Scotland and *Mercury* flew on to complete 6,045 miles non-stop to the Orange River in South Africa, setting a distance record for marine aircraft which has only in recent years been surpassed by a Russian Beriev Be–12 amphibian and which has never been equalled by a seaplane.

Imperial's other initiative was to take up an idea of Sir Alan Cobham's for inflight refuelling of flying boats. Again, the objective was to overcome the loss of a substantial volume of fuel during take-off by, in effect, 'topping up' once the aircraft was airborne, thus

The brilliant Short Mayo composite aircraft, with *Maia* ready to give *Mercury* a flying-start!

Off the south coast, spring 1939, an Armstrong Whitworth A.W.23 bomber refuels Imperial Airways' Short Empire flying boat *Cambria*.

extending the range of the flying boat. Early experiments used an elderly Armstrong-Whitworth A.W.23 bomber modified to act as a flying tanker. This refuelled *Cambria*, in spring, 1939, during a series of experiments based upon Ford Aerodrome in Sussex. A series of eight transatlantic flights were made during August, 1939, in the four weeks before the outbreak of World War II in Europe. On the transatlantic flights, Handley Page Harrow II bombers of Sir Alan Cobham's company, Flight Refuelling Ltd., refuelled Imperial Airways flying boats over Foynes in Ireland and over Botswood, Newfoundland, on flights between Southampton and Montreal, starting on 5th August. The series of flights passed without incident.

Meanwhile, during the summer months of 1938 and 1939, both airlines operated a number of return transatlantic flights. After considering aircraft proposed by Sikorsky and Martin, and a novel five-engined twin-fuselage flying boat from Seversky, Pan American had selected the Boeing Model 314, a double-deck monoplane flying boat with four 1,600-hp Wright Twin Cyclone radial engines, able to carry up to seventy-four passengers over 3,500 miles. This large flying boat had a wingspan of 152 feet and was 106 feet long, and used

A Flight Refuelling Handley Page Harrow II refuels a Short Empire at the start of a transatlantic crossing.

sponsons rather than wingtip floats. There was a passageway inside the wing, running direct from the back of the flight deck, so that the engineer could check the engines during flight. The flight deck was 21 feet long and 9 feet wide, with a spiral staircase down to the passenger deck, a feature to be found on the Boeing 747 today.

The prototype 314 made its first flight on 7th June, 1938, staying in the air for thirty-eight minutes, although difficulties during take-off and landing led to a lowering of the position of the sponsons, and the original single fin was replaced by a triple fin. Six of these aircraft were built for Pan American, including the first test-flight machine, and these were delivered between January and June, 1939, with the names *Honolulu*, *California*, *Yankee*, *Atlantic*, *Dixie* and *American Clipper*. On 20th May, 1939, *Yankee Clipper* flew from Port Washington, New York, to Lisbon, Portugal, in just under twenty-seven hours. A little more than a month later, on 24th June, this aircraft made the type's first return flight to Britain, and four days later *Dixie Clipper* made an experimental passenger flight to Britain, while Imperial flew from Poole to Baltimore. However, war was looming, and the operation was, after all, only seasonal, so while *Cabot* made a record run of thirteen hours westbound on 1st October, by 30th October, when *Dixie Clipper* made the last eastbound flight to Bermuda and Lisbon, she carried just one passenger, who got off at Bermuda.

A further six Boeing 314s were built, three for Pan American for Pacific operations, and three for the British Overseas Airways Corporation, the successor to Imperial Airways, while BOAC's own long-range flying boats, the 'G' class, served with the RAF.

Deutsche Luft Hansa, meanwhile, had received the first large, four 600-hp Junkers Jumo diesel-powered Blohm und Voss Ha.139 gull-wing seaplane in March, 1937, and this aircraft, *Nordmeer* or North

The ultimate in Boeing and Pan American flying boats, the giant Boeing 314.

Sea, was followed by *Nordwind* or North Wind. Both aircraft operated from DLH's launching ships, but in this case operations took place over the North Atlantic rather than the South Atlantic, with the *Schwabenland* based off the Azores and another ship, the *Friesenland*, operating off Long Island, New York; between August, 1937 and November, 1937, these aircraft made seven experimental transatlantic return trips, with a 1,000 lb payload on each flight. There were some problems with engine overheating, but generally the aircraft performed well, and a third aircraft, *Nordstar* or North Star, joined DLH in 1938. During 1938 and 1939, there were further flights across the North Atlantic between Horta and New York, and across the South Atlantic between Bathurst and Natal or Recife in Brazil, with a total of forty North Atlantic and sixty South Atlantic flights being completed without incident by June, 1939.

The same engines, and again with four to each aircraft, were also used by the last Dornier flying boat, the Do.26, although in two nacelles and with the aft engine in each nacelle being 10 degrees out of

Luxury travel meant having separate dining-rooms rather than meals in a cramped seat – aboard a Pan American Boeing 314.

the horizontal to keep the propeller clear of the water during take-off and landing – a useful point since flying boats need to adopt a nose-up stance on the water to avoid porpoising. The traditional Dornier sponsons on this aircraft were replaced by retractable wingtip floats. Three Dornier Do.26s were built, *Seeadler*, *Seefalk* and *Seemowe*, with the first aircraft making the first flight on 21st May, 1938. Again, these aircraft operated on the transatlantic services, but only for a short period before war broke out and, with the Ha.139s and three Do.26Cs with uprated engines, all passed into Luftwaffe service. A similar fate awaited Germany's large transatlantic flying boat, the six-engined Blohm und Voss Bv.222 Wiking, or Viking.

In France, there was the same urge for aircraft with the capability of spanning the oceans, although landplanes featured more prominently in operations to the French African possessions, leaving the flying boat for operations to Indo-China and for French ambitions on the Atlantic. One of the first French designs for a large flying boat was the Latécoère 300 series, first flown in 1931 with four 650-hp Hispano-Suiza engines mounted in two tandem nacelles above the monoplane mainplane. Intended for the air mail service between Dakar and Natal, the one and only Latécoère 300 sank in December, 1931, at Marseilles. However, it was salvaged and ready to be flown again on 7th October, 1932, being given the name *Croix du Sud* or Southern Cross. On 31st December, 1933, this aircraft flew from a lake near Berne to St Louis in Senegal in just twenty-three hours non-stop – a distance of 2,300 miles. The following year the aircraft made a crossing between Dakar and Natal on 3rd and 4th January. Unfortunately, the aircraft was lost in mid-Atlantic on 7th December, 1936, after further attempts at transatlantic flight. However, belatedly impressed by the design, the French Government ordered six aircraft, three Latécoère 301s for the Armée de l'Air and three Latécoère 302s for the Aéronavale, with the first 301 flying on 7th September, 1935. Latécoère aircraft seemed to be fated, with the Latécoère 521, a 162-foot wingspan giant monoplane with sponsons and six Hispano-Suiza engines, flying for the first time in January, 1935, but being wrecked at its moorings in 1936. However, after rebuilding, this aircraft set several payload-to-height records, proving a worthy rival to the Sikorsky S–42, although the French aircraft was intended to carry no more than thirty passengers. Even more dramatic, the 631, on which work started in 1937, had its prototype nearing completion as France fell in May, 1940. The aircraft was dismantled and hidden in the Forest of Landes, but after liberation in December, 1944, it was reassembled and flew again the following year. Two of these six-engined monoplanes were built,

The one and only *Gooney Bird*, the floatplane conversion of the Douglas C–47/ DC–3 Dakota.

operating on longer overseas services and carrying up to eighty passengers.

Of course, these were only the flying boats and seaplanes designed as such. Seaplane versions of transport aircraft, especially of the smaller types, were not unknown, and indeed at the lower end of the scale such conversions were commonplace and essential. Several Junkers Ju52/3M tri-motors were built as seaplanes, especially for airlines such as Finnair, while a single Douglas DC–3/C–47 Dakota, known as the *Gooney Bird*, was also built as a seaplane. Yet, while the lack of suitable landing fields made such conversions worthwhile, their future and that of the purpose-designed flying boat was already under threat, with airlines such as Pan American actively searching for landing fields for landplanes on the North Atlantic route as war broke out, and Imperial Airways having built an airfield at Ta Kali in Malta during the mid-1930s. The lower weight, greater streamlining and high speed of the landplane meant that as the 1930s closed, it was already becoming a serious competitor to the more staid flying boat.

6
ON PATROL

To place a date on the switch from biplane to monoplane is impossible. Obviously the change did not occur overnight, if only because many of the old biplanes had plenty of life left in them and, as still happens today, there is sometimes relatively little in performance between a late mark of an older design and the early versions of a newer one. Conservatism in some air arms kept the biplane on the production line even while others elsewhere had introduced monoplanes to their squadrons. Nor was the monoplane new, for after all, the Schneider Trophy had been won by a monoplane in 1913. However, the difference between the early monoplane and the aircraft which came into prominence during the 1930s was the lack of external bracing, made possible by the development of the cantilever wing during the late 1920s; the amount of bracing required on some of the earliest monoplanes created substantial drag, even compared with some biplane designs. For shipboard use, the extra lift and reduced wingspan permitted by the biplane ensured its survival into World War II and beyond. There were even a few landplanes, such as the de Havilland Rapide and Moth, which remained in production into the early post-World War II period, and yet were biplanes from the same design office as that which designed such excellent monoplanes as the Mosquito and Sea Hornet.

However, by the early 1930s, the monoplane was definitely taking over for most purposes, and aircraft were also becoming sleeker in many other ways, with considerable attention being paid to streamlining: landplanes were being built with retractable undercarriages; bombers were having their gun positions enclosed to match cockpits which had already been enclosed; engines were being cowled and faired into wings; and tailplanes were becoming simpler, but more robust. The idea of cowling over engines to reduce drag had come from the National Advisory Committee on Aeronautics, or NACA, so these were usually termed as 'NACA cowls', and the organization was effectively the predecessor of NASA, the National Aeronautics and Space Administration.

The generation of aircraft developed during the 1930s was of special significance in other ways, for often these were either the aircraft to be used during World War II or, if not, the aircraft of the war were direct descendants of them. German rearmament occurred during the 1930s, allowing German designers to see their aircraft built in their own country, while in Japan, boosted and inspired by examples of British and American technology, the growing Japanese aircraft industry was sponsored by the armed forces and encouraged to build a wide range of aircraft with which Japan could expand its territorial ambitions. In an unsettled world, preparing for war, the flying boat, amphibian and seaplane could not help but play a part.

Although faster aircraft, and of course that meant fighters, were generally landplanes, with the development of the aircraft-carrier and the growing size of these vessels obviously encouraging this trend, a rearguard action was fought by Seversky, a Russian *émigré* designer who enjoyed some success during the early 1930s in the United States. The logic was simple and irrefutable, for not everywhere offered a concrete runway, and aircraft-carriers were still unknown in most of the world's navies, with even the United States then lagging behind Britain in their development. Then too, the world air speed record had only recently been established by a seaplane! In fact, Alexander de Seversky built a monoplane amphibian fighter, which he designated the SEV–3M–WW and the United States Army Air Force designated the P–35, and this aircraft established an amphibian speed record in 1933. A three-seat enclosed cockpit meant that an already heavy aircraft was still heavier, and the single 440-hp Wright radial engine provided a maximum speed of just 175 mph on the standard production aircraft. However, all of the production aircraft ordered by the USAAC were completed as landplanes. Undaunted, de Seversky built as a private venture, an amphibian which he called the 'Convoy Fighter', and the Imperial Japanese Navy bought no less than twenty, giving the aircraft the doubtful distinction of being the only standard production American aircraft to be used operationally by Japanese forces prior to the rehabilitation and rearmament of Japan during the 1950s. Ironically, given Seversky's background, a development, the 2PA–A amphibian, was obtained by the USSR in 1938.

It was still not unusual for aircraft to be built in both landplane and seaplane or floatplane versions, like the Fairey Swordfish biplane for instance, although invariably aircraft were becoming more specialized as operating requirements became more demanding in terms of speed, range and armament capacity. This reduced the designers' freedom to design aircraft which could be maids of all work, and to achieve the optimum performance landplanes, seaplanes, flying boats and amphibians were becoming designed as such.

The decline of the seaplane was matched by a reduction in the

Against the spread of the aircraft-carrier, the seaplane-carrier was a dying breed; this was HMS *Albatross* after her transfer from the Royal Navy.

number of seaplane-carriers, or tenders as the United States Navy preferred to call them. The Royal Navy maintained just one such ship during the post-World War I period, HMS *Albatross*, and this ship was withdrawn in 1929 and then transferred to the Royal Australian Navy, which operated the ship until 1933, although she returned to service during World War II. In Royal Australian Navy service, the aircraft operated were mainly from the Supermarine Seagull series. The French Navy operated the *Commandant Teste* seaplane-carrier, with Lioré 210 fighters and Latécoère 298 torpedo-bomber seaplanes, but even there the seaplane-carrier took second place to the aircraft-carrier and its landplanes, for while the Marine Nationale operated just one carrier between the wars, the converted battleship *Béarn*, two carriers were being built at the time of the fall of France, the *Joffre* and the *Painleve*. Only Japan had long-term plans for the seaplane, building the world's most advanced and fastest seaplane-carriers, the *Chitose* and *Chiyoda*, which entered service in July and December 1938 respectively. These could each carry up to twenty-four seaplanes, could operate at just below 30 knots, and had an endurance far in excess of that of any other class of Japanese warship. Both ships had four catapults and one stern and six amidships cranes. However, they

One of the finest seaplane-carriers ever, the Japanese *Chitose* before conversion to an aircraft-carrier.

were converted after the outbreak of war to become midget submarine tenders and after that aircraft-carriers.

But, in spite of all this progress, seaplanes were still needed for shipboard operations, and there was a difference between the American and German practice, in using pure seaplanes for fleet-spotting use, and the British and Australian practice, which was to use amphibians.

Abandoning its policy of designing aircraft and putting the contracts out to tender, BuAer sought designs from American manufacturers for a lightweight fleet-observation seaplane during the early 1930s. The three main contenders for this contract were the Curtiss XO3C–1, the Douglas XO2D–1 and the Vought XO5U–1, all of which were powered by a single Pratt and Whitney engine and bore a further resemblance to each other in having a single float and using a monoplane configuration. In prototype form, the aircraft were completed as amphibians. Two 'outsiders' were also considered, the Bellanca XSOE–1, which used a 725-hp Wright engine, and the Fairchild XSOK–1, based on a successful civil flying boat. The Bellanca eventually proved to be overweight, and never flew. The three main contenders had approximately the same weight and performance, as one might expect with so many other similarities, and typically, the XO3C–1 gave a maximum speed of 159 mph with an all-up weight of 5,236 lbs on its first flight at Buffalo on 5th March, 1934.

The Curtiss aircraft won the competition, and the production aircraft, designated the SOC–1 Seagull, were built as seaplanes with enclosed cockpits, the first two aircraft joining the cruiser USS *Marblehead*. There were 135 SOC–1 Seagulls in the first production batch. A basically similar aircraft, the SOC–2, entered production in 1936, after deliveries of the SOC–1 had been completed. There were forty of this version, while a further development, the SOC–3, included eighty-three aircraft for the USN and USCG built by Curtiss. Another forty-four were built at the Naval Aircraft Factory as SON–1s, in 1937, for the USN.

The Naval Aircraft Factory also produced a design of its own, the XOSN–1 which first flew in May, 1938, to meet a USN requirement for an aircraft that could operate from battleships or carriers without the need for folding wings. A private design, the X–85 or XOSS–1, for the same USN requirement, was produced by the light aircraft manufacturer Stearman. However, the all-metal Vought OS2U–1 Kingfisher was the preferred aircraft, using a 450-hp Pratt and Whitney engine, and first flew as a landplane on 1st March, 1938, and as a seaplane on 19th May, 1938. The initial order for fifty-four OS2U–1 Kingfishers was followed by a further 158 OS2U–2s, which had leak-proof fuel tanks, although many of these aircraft were built as landplanes with floats available for conversion to seaplanes when necessary. The main production version of the Kingfisher was the OS2U–3, of which 175 seaplanes and 831 landplanes, again available with floats for rapid conversion to seaplanes, were built, including a hundred aircraft for the Royal Navy to operate from merchant cruisers during World War II. Deliveries of these aircraft took place throughout the early years of the war.

All of these aircraft were intended for reconnaissance and fleet-spotter duties, acting as the eyes for the big guns of the battleships and cruisers in a role not unlike that of *Engadine*'s aircraft during World War I. The British approach tended to be slightly different from that of the Americans, developing a more up-to-date version of the Supermarine Seagull series. Initially the new aircraft was known as the Seagull V, but before long this was changed to the Walrus, one of the most popular marine aircraft of World War II. First flown on 21st June, 1933, using a single inter-wing 635-hp Bristol Pegasus II radial engine to drive a pusher propeller, the Walrus differed from American practice in being a biplane and an amphibian. The initial orders were

Typical of American thinking on observation seaplanes during the 1930s and 1940s was the Stearman XOSS–1, designed for a BuAer competition in 1938, but unsuccessful.

for twenty-four aircraft for the Royal Australian Navy and just twelve for the Royal Navy. Before long, further orders were given, totalling 204 aircraft, of which three were delivered to the Irish Army Air Corps. At first, the metal-hulled Walrus I was intended for Fleet Air Arm operation from battleships and cruisers, with the first aircraft entering Fleet Air Arm service during the summer of 1936, initially with 700 squadron, and then later with 711, 712 and 714 squadrons – all in the Fleet Air Arm's 700 series of utility or support squadrons – combat squadrons are in the 800 series. Although basically a reconnaissance aircraft, the Walrus did have some teeth, with two machine-gun positions, one in the bow and the other amidships, with 0.303 Vickers machine-guns, while a small load of bombs or depth-charges could be carried on under-wing racks. However, the aircraft soon found itself with another role, and doubtless the one which earned it such popularity during the war years – taking on vital search and rescue duties for the Royal Air Force and the Fleet Air Arm. This released larger Sunderland and Catalina flying boats for longer-range duties, including maritime reconnaissance, while downed airmen in the English Channel and the North Sea were picked up by the Walrus amphibians. An uprated version, the Walrus II using Pegasus VI engines, appeared during the war years but, due to a shortage of materials, many of these were built with wooden hulls by Saunders–Roe at Cowes, thereby releasing Supermarine to build other more urgently needed aircraft, including the landplane development of the S.6B seaplane, the Spitfire fighter. Production of the Walrus ended in 1944, but this small aircraft, with a wingspan of just 45 feet and 37½ feet long, and with a maximum speed of 135 mph, remained in service for many years afterwards, with the Argentine Navy retaining its aircraft until 1966. However, in the British and Australian services the aircraft was soon replaced by the Sea Otter, another biplane amphibian, and then by the helicopter. By that time, of course, the reconnaissance rôle of the shipboard aircraft had been reduced by the advent of radar, and major fleet actions were no longer between the guns of opposing battle fleets, but the province of carrier-borne bomber aircraft.

Not all amphibians were shipborne and, during the mid-1930s, Douglas built a number of amphibians for civil, military and naval use, with two engines mounted above a cantilever monoplane wing. Essentially similar, the USN's XP3D–1 used Pratt and Whitney Wasp engines, and the USAAC's YOA–5 used two 930-hp Wright Cyclones, the latter variant also bearing the designations YB–11 and YO–44 during its development stages. The first YOA–5s were delivered on 24th February, 1935, for search and rescue operations, while the XP3D–1 was redesignated as the XP3D–2 in May, 1936, with the then novel feature of retractable wingtip floats. But even so, this naval version, with its metal hull and metal leading edges to the mainplane, was not put into production since the Convair Catalina offered a superior performance. However, the USSR and Japan ordered the aircraft, and both of these countries received two machines described by the manufacturer as the Douglas 'DF'.

Smaller, and perhaps closer to the Walrus concept, was one of the early Grumman amphibians, the G–21A or, in USN service, the JRF, although more usually known by its civil title of Goose. The first Grumman amphibian had appeared some four years before the Goose, and this aircraft, a utility biplane amphibian with a large single float, known as the XFJ–1 Duck, had first flown on 24th April, 1933. A single 700-hp Pratt and Whitney Twin Wasp had provided the then fairly high speed for an amphibian of 170 mph, and the aircraft could carry a single gun, fired from the rear cockpit, if necessary. By contrast, the Goose was a twin-engined monoplane of 49 foot wingspan

The J2F–4, one of the Duck series of the early 1930s – Grumman's only single-engined amphibian design.

and 38½ feet in length, and the prototype, which first flew in June, 1937, used two 450-hp Pratt and Whitney Wasp Junior engines. The first customers were civilian and commercial operators, but a single aircraft was acquired by the United States Navy in 1938, designating this as the XJ3F–1 general utility aircraft, before later adopting the JRF designation. Orders followed what must have been a successful service evaluation, for ten JRF–1s and ten JRF–4s, with the latter able to carry two 250-lb bombs or depth-charges on under-wing racks, and these aircraft were delivered in 1939. The USAAC also ordered Goose amphibians in 1938, designating these as OA–9s and ordering twenty-six machines initially, while a further five civilian aircraft were conscripted into the USAAC after the entry of the United States into World War II. Not to be outdone, the USCG ordered ten aircraft for

Conforming more closely to the standard Grumman pattern was the JRF–1 Goose, which saw service in World War II.

search and rescue duties during 1939 and 1940. The most popular variant of the Goose series was the JRF–5, of which two hundred were built, with a range of up to 640 miles and a maximum speed of 200 mph. Inevitably, a small number found themselves in RAF service, although the original fifty ordered by the British Purchasing Commission in 1940 were whittled down to six by the wartime demands of the USN and USCG. Other aircraft found their way, during World War II and afterwards, into the Royal Canadian Air Force, the Portuguese Navy and French Navy, or Aéronavale, service.

Meanwhile, while the British were still depending upon updated versions of their Blackburn Iris and Supermarine Southampton flying boats, notably the Blackburn Perth, the Italians were showing that even older designs could still be used effectively: Balbo was back again, with an updated S.M.55.

Balbo used the final military version of the S.M.55, the S.M.55X, with 750-hp Asso engines and a stretched fuselage, which had a maximum speed of 165 mph and a range of 1,250 miles. His last formation flight with the S.M.55 series was to be most spectacular, with twenty-four aircraft flying in formation from Rome to Chicago to arrive in time for the 1933 Chicago World's Fair. The formation left Orbatello, near Rome, on 1st July, 1933. The twenty-four aircraft flew in flights of three aircraft, each in a 'V' shape, and took 7¼ hours for the 875-mile flight from Rome to Amsterdam, over the Alps: clearly getting there together was far more important than speed. At Amsterdam, tragedy struck. One S.M.55X hit a mud bank on

For his ultimate formation flight, Balbo used the Savoia–Marchetti S.M.55X twin-hulled flying boat.

landing and capsized with the loss of one crew member and injury to another. Determined to maintain a twenty-four aircraft formation, Balbo substituted one of two reserve aircraft. The North Sea was crossed in heavy rain, and the formation flew on over England to refuel at Londonderry in Northern Ireland. It remained there for three days while bad weather passed before attempting the exposed 950-mile flight to Reykjavik in Iceland, there being no alternative landing sites between these two points. Even so, most of the flight to Iceland was in fog, except for the landing, which took place in heavy rain! Further delays occurred due to bad weather, before the formation eventually reached Cartwright in Labrador, and then flew to Shediac Bay in New Brunswick, before flying on to Montreal and landing in the St Lawrence. Further mishaps were narrowly avoided at Montreal, with many of the aircraft having to swerve while landing, trying to avoid over-enthusiastic and ill-disciplined spectators in small boats. One account of the landing tells of an enthusiastic radio reporter putting his microphone near to Balbo, who had landed first and was watching the rest of his formation attempt to avoid the small boats, so that his listeners could hear what he thought was the Italian's excitement at the enthusiastic welcome – but those who understood Italian were treated to a choice example of Italian obscenities! After an overnight stop at Montreal, the formation took off for Chicago the following day, carrying the Italian Ambassador to the United States for this final stage, and as they arrived at Chicago on 15th July, the USAAC's First Pursuit Wing, based at nearby Selfridge Field, provided a fighter escort. The formation flight had covered 11,500 miles in 48 hours, 47 minutes, at an average speed of 124 mph and the loss of one aircraft and one airman.

A few days later, on 19th July, Balbo flew from Chicago to New York and Washington, lunching with President Roosevelt in Washington, before returning to New York for a ticker-tape parade. On his return to Italy, Balbo's formation took a more southerly route back through the Azores, and on taking off from the Azores, on 8th August, one aircraft crashed, injuring three crewmen slightly, but the pilot, Lieutenant Enrico Squaglia, died later from shock. Flying to Rome via Lisbon, Balbo eventually arrived back at his base on 12th August, 1933, to be greeted by an escort of fifty flying boats and seaplanes of the Regia Aeronautica (RA) and by Il Duce, Benito Mussolini, himself. Balbo's award for these achievements was to become Italy's first air marshal, and later Governor of Libya – although the latter position suggests that he was wanted out of the way after the public acclaim had died down and, rather oddly, he was later shot down while piloting an aircraft by his own anti-aircraft fire after failing to identify himself!

The S.M.55 remained in service into the early years of World War II, although a development, the civil S.M.66, remained in service throughout the war years, operating as a search and rescue flying boat and as a military transport, having been conscripted into RA service.

Savoia–Marchetti was not the only significant flying boat builder in Italy during the 1930s, for apart from Piaggio, there was a new arrival in the shipbuilding firm Cantieri Navale Triestino, which started seaplane and flying boat production as Cantieri Rivniti dell'Adriatico (CRDA). The first of a series of flying boats was the CRDA Cant. Z.501 (Z for Zappata, the designer), with a single 900-hp Isotta–Faschini Asso engine faired into its 73¾-foot wingspan monoplane wing. This aircraft had the not inconsiderable speed of 170 mph, and was intended for long-range maritime reconnaissance and bombing operations, being given the name Gabbiano or Seagull – although it did not have a seagull wing – after its first flight at Trieste in the summer of 1934. Later that same year, in October, the Gabbiano established a world distance record of 2,560 miles, flying non-stop from Trieste to Massawa in Eritrea, and although this record was of short duration, the aircraft re-established its rights to a distance record in July, 1935, flying 3,080 miles from Trieste to Serbera in Somaliland. Production Z.501 Gabbianos entered the Regia Aeronautica service in 1936, and by the outbreak of World War II the RA had some two hundred Z.501s in twenty reconnaissance squadrons. The type was still in service at the time of Italy's surrender in 1943. Even at this late date, the Z.501 was an all-wood flying boat; fitted with three 7.7-mm Breda machine-guns in three positions, it could lift up to 1,400 lbs in bombs on under-wing racks. The aircraft saw service in the Spanish Civil War with the Nationalist forces, and in 1937 and 1938 a number were sold to Romania.

Another Zappata design for CRDA which also saw service in the Spanish Civil War was the Z.506 seaplane, a twin-float design with three engines, usually Pratt and Whitney Hornet radials, although on some aircraft Wright Cyclone engines were used, or even Alfa Romeo engines, all usually of 750 or 800 hp. Unlike the purely military Z.501, the Z.506 was available in commercial versions, but the aircraft also followed the record-setting achievements of the Z.501, establishing several records for distance-with-payload and speed in the seaplane classes. Primarily a maritime reconnaissance and bomber aircraft, the Z.506 in its military form was designated the Z.506B and named the Airone or Heron, joining the Regia Aeronautica in 1937. It was capable of lifting up to 2,200 lbs of bombs, and was fitted with a ventral machine-gun turret below the fuselage and an amidships turret, for a single 7.7-mm Breda. Production outstripped CRDA's facilities, and in addition to being built at its home factory, the aircraft was also produced by Piaggio. It served throughout the war years, with a small number remaining in search and rescue and light transport units until 1959. The only export customer was Poland, but these aircraft were never delivered because Russia and Germany invaded Poland before deliveries could commence.

One of the first German seaplanes to be built in Germany after the Versailles Treaty at the end of World War I was a twin-float biplane, the Heinkel He.59, with two inter-wing 660-hp BMW engines mounted on either side of the fuselage. Like the Z.506, this aircraft also had a gun position under the fuselage as well as one behind the cockpit and another in the nose. Originally intended to be either a landplane or a seaplane, only aircraft of the latter type were built after the first flight of the landplane form in September, 1931, although it did not fly as a seaplane until the following January. The plane's similarities with the Z.506 extended to a similar armament of up to 2,200 lbs of bombs and service in the Spanish Civil War, although the earlier He.59 was regarded by the Luftwaffe as obsolete even before the outbreak of World War II and was relegated to search and rescue duties or to training units. Many He.59s were built in Germany by Arado, who also built the Arado Ar.196 seaplane for the German Navy's battleships and cruisers. Another smaller, biplane seaplane for shipboard use was the Heinkel He.60, a development of the He.45 and He.51 landplanes. The He.60 was also used as a coastal reconnaissance aircraft, and aircraft of this type flew in the Spanish Civil War on the Nationalist side. Heinkel also built the successor to the He.59, the He.115, which was sold to Norway and to Sweden.

Generous financial incentives also encouraged many German companies to enter aircraft production, joining the well-established companies such as Junkers, Heinkel and Dornier. One of the most notable of these newcomers was the shipbuilding firm of Blohm und Voss, which had established a factory at Hamburg in 1933. One of the first Blohm und Voss designs was the Bv.138 flying boat, a three-engined twin-boom fuselage flying boat for reconnaissance duties, designed for shipboard use and first flown in 1937. The Bv.138 required a very substantial stern-mounted launching catapult. The Bv.138 was used for shipboard reconnaissance, shore-based maritime reconnaissance and for mine counter-measure duties, with a large detector ring to detonate mines. Later Blohm und Voss designs

Designed for Luft Hansa, the Dornier Do.18 joined the Luftwaffe in 1938 as a search and rescue machine.

included a large four-engined seaplane with a monoplane 'gull' wing and four 600-hp Junkers Jumo diesel engines and capable of up to 200 mph. This was the Ha.139, designed for Luft Hansa for Atlantic mail flights, and oddly this large and fast aircraft was not built for military use, even though a maritime reconnaissance version was seriously considered. Later, however, Blohm und Voss was to build Germany's first and largest World War II aircraft, the Wiking flying boat.

By the mid-1930s, Dornier had lost the urge to build ever larger flying boats and had acquired both the taste and the reputation for building some of the most attractive flying boats and seaplanes, including the Dornier Do.18, Do.24 and Do.26. Both the Do.18 and Do.26 were designed with the needs of Deutsche Luft Hansa in mind, as we saw in the last chapter. A successor to the Wal series, the Do.18 had an enlarged and streamlined configuration vaguely reminiscent of the earlier aircraft, with twin 880-hp Junkers Jumo diesels in a single nacelle faired in to the wing, and of course with both pusher and tractor propellers. This aircraft, with a span of 78 feet and 63 feet in length, could fly at up to 165 mph and had a range of 2,175 miles. The first flight was on 15th March, 1935, although the aircraft did not 'join'

Similar to the Blackburn Sydney, the Dornier Do.24K was built by Aviolanda for the Royal Netherlands Navy.

the Luftwaffe until 1938, as the Do.18D and Do.18G, equipped with machine-guns in the nose and midship positions. The type was only used on search and rescue missions.

Still more streamlined was the Do.24, the aircraft designed originally to fit the needs of the Royal Netherlands Naval Air Service in the East Indies, replacing elderly Wal flying boats. Larger than either the Wal or the Do.18, the Do.24 had a wingspan of 88 feet and was 72 feet long, yet was capable of 210 mph and had a range of 1,800 miles using three 1,000-hp BMW Bramo radials. However, for the first flight, 3rd July, 1937, the prototype used 600-hp Junkers Jumo diesels. The RNAS aircraft, designated the Do.24K, also used the Cyclones and were built in the Netherlands by Aviolanda and de Schelde. After the fall of France and the Netherlands, the Luftwaffe received 150 Dutch-built and fifty French-built Do.24Ns. Post-war, the type continued to serve with the French Aéronavale and with the Spanish Air Force. The four-engined Dornier Do.26 was designed for Luft Hansa.

During the years before war broke out in Europe, Fokker in the Netherlands designed a torpedo-carrying and reconnaissance

Another wartime conscript from DLH to the Luftwaffe was the Dornier Do.26.

seaplane, a twin-float aircraft, with twin 450-hp Wright Whirlwind engines, similar to the Heinkel He.115 and known as the T.VIII-W. This aircraft first flew in 1939, was capable of 177 mph and had a range of about 1,300 miles. An initial five T.VIII-Ws were ordered for the Marine Luchtvaardienst, the Dutch Navy's air arm, and these were followed by fourteen T.VIII-W/Gs, the G suffix referring to *gemengd*, or 'mixed', a reference to the mixed wood, metal and fabric construction. Several T.VIII-W/Gs passed into Luftwaffe service after the fall of the Netherlands, followed by a number of T.VIII-W/Ms, all-metal versions, built after the occupation. A small number of aircraft escaped to France, and then to Britain, with their crews.

Before the outbreak of World War II, the Luftwaffe, under the strong political leadership of its head, Hermann Göring, acquired the right to operate all German service aviation, even that in support of the Army, using Fiesler Storch light aircraft and the German Navy's Arado Ar.196 fleet reconnaissance seaplanes. The Navy in particular was to suffer from this change, which was in due course to delay and finally force the abandonment of plans for German aircraft-carriers, inhibiting the German war effort at sea.

In France, the Aéronavale retained control of its aircraft, introducing the Lioré 70, a three-engined 'hydravion de surveillance catapultable', for shipboard use, in spite of a wingspan of 98 feet and a flying boat hull: perhaps for this reason, only eight were built, and of these six were allocated to Escadrille (squadron) E.7. Unusually, the aircraft had two tractor and one pusher propeller. The French continued to suffer problems in funding adequate procurement programmes throughout the 1930s, and it was not until the outbreak of war that a number of new and high-performance French flying boats and seaplanes appeared – too late! One small French single-engined twin-float seaplane that did achieve production in volume, however, was the Latécoère 298, which used an 880-hp Hispano-Suiza engine, and which could manage 180 mph and a range of 500 miles. Used for torpedo-bomber or reconnaissance operations, the first flight of the prototype was on 8th May, 1936, and this was followed by trials with the Aéronavale. The initial order was for thirty-six aircraft, all of which had the rather old-fashioned feature of a propeller-synchronized 20-mm cannon firing through the propeller disc in addition to wing-mounted Darne machine-guns. A number of these aircraft were built with folding wings for operation from the *Commandant Teste* seaplane-carrier and were designated Laté 298B. Altogether, some sixty aircraft were in service by the fall of France, serving with six squadrons, although in the short period between the beginning of the German invasion and the fall, these aircraft were operated as dive-bombers. Aircraft surviving the German onslaught mainly passed to the Vichy Government, although a number of aircraft were taken by their pilots to join the Royal Air Force. Post-war, a small number remained in use as trainers until 1950.

A comparative wasteland for marine aircraft production at this time was the Soviet Union, perhaps not so surprisingly for a continental power with most of its coastline, in the Arctic or on the Baltic, frozen for much of the year. The political and economic isolation of post-revolutionary Russia did little for international air links, and nor did the extreme poverty of the country. Few Soviet flying boats of the inter-war years stand out, or are even worthy to be compared with the efforts of designers in Britain, Europe and the United States, and indeed the homeland of Sikorsky and de Seversky lagged behind in aeronautical development generally at this time and indeed until the post-World War II period. However, one could understand the desire of great designers to be free of the Soviet system, for apart from its oppressive effect, other designers who remained in the USSR became the victims of political purges, and Tupolev did much of his early work while in a prison camp! Motivation of talented scientists and designers hardly seemed to be a strong point of the revolutionaries.

One of the few successful Soviet flying boat designers, before, during and after World War II, was Georgi M. Beriev, designer of the MBR, a short-range maritime reconnaissance flying boat, which first flew in prototype form during 1931, using a single 500-hp BMW, mounted above the monoplane mainplane and driving a pusher propeller. An uprated engine, also obtained from BMW but offering 680-hp, was used for the early MBR-2 production aircraft, with the engine eventually built under licence in the Soviet Union. Deliveries of the production aircraft started during 1935. Despite the MBR–2 being a 1930s design, the aircrew still sat in the open, including the pilots and the gunners in the bow and in the mid-upper gun position, with their 7.6-mm machine-guns. Although the aircraft could be modified for land operations with the addition of a fixed undercarriage, it was not an amphibian, and other modifications included fitting skis for take-off from snow or ice. A major redesign in 1935 created the MBR–2bis and included some aerodynamic modifications and the substitution of enclosed accommodation for the crew. The aircraft remained in production throughout World War II, although the maximum speed was just 170 mph and the range 400 miles.

Rather larger than the MBR–2 was the MDR–3 four-engined flying boat, which is believed to have made its first flight in 1932. Designed

Soviet designers were not very successful during the late 1930s, with one of the rare Soviet flying boat designs being the MBR–2 – this was a later MBR–2bis with rounded tail and enclosed gun positions.

by Igor Chetverikov, the aircraft required the attentions of a number of other designers but, even so, failed to overcome its basic design faults and only a small number were built. Chetverikov's work seemed to be doomed to failure, for although his ARK–3, a reconnaissance and utility flying boat, with twin 710-hp M–25 radial engines mounted in tandem in a single nacelle above the mainplane, achieved an altitude record of 30,151 feet on 25th April, 1937, the small number of ARK–3s in Soviet Navy service were soon withdrawn due to structural failure. A later Chetverikov flying boat, the MDR–6, first flown in 1937, appeared to be more successful, and fifty aircraft were built.

By contrast, in the United States there was a never ceasing search for bigger and better flying boats and amphibians, as well as seaplanes. Amongst the wide variety of designs available for consideration by the United States Navy, and its guide, and mentor, BuAer, were aircraft such as the Hall XPTBH–2 twin-float seaplane, a monoplane with two engines, built for patrol, torpedo and bomber operations, which was truly comparable with the Heinkel He.115. Originally designed as the XPTBH–1, a switch from Wright Cyclone engines to twin 800-hp Pratt and Whitney Wasps on the aircraft delivered to the USN for evaluation on 30th January, 1937, resulted in a slight change of designation to XPTBH–2. The aircraft could manage more than 180 mph. Another twin-float patrol seaplane was the Northrop N–3PB, a rather faster aircraft with a maximum speed of 250 mph, and which used a single 1,200-hp Pratt and Whitney Wasp, but this aircraft was not ordered for the United States Navy. Nevertheless twenty-four N–3PBs were ordered by the Royal Norwegian Air Force, and these were eventually delivered in 1940, after the fall of Norway, to be operated by Norwegians in exile as No.330 Norwegian Squadron Royal Air Force, on anti-submarine patrols.

Competition arose between Sikorsky and Consolidated during the mid-1930s as both companies sought the prize of building a new long-range flying boat for the United States Navy. Competition between two such companies could not fail to provide a worthy aircraft, and it did. Consolidated, in effect, had an edge on Sikorsky, for while the latter company had done well in dominating the commercial flying boat and amphibian market for several years Consolidated was more attuned to the needs of the United States Navy. Consolidated had not had it all its own way, however, having seen its XPY–1 of 1929 rejected by the USN in favour of aircraft built by Glenn Martin, even though the design had resulted in orders for a commercial version for use by Pan American Airways on the airline's Caribbean services. Success with the USN finally came for Consolidated in 1932, with the P2Y–1, a twin-engined sesquiplane development of the XPY–1, and this aircraft, with its 100-foot wingspan, 62-foot-long fuselage, 125-mph cruising speed and 2,000-mile range, proved to be the company's salvation during the worst of the Depression. The aircraft joined the USN, initially with Patrol Squadron 10, or VP–10, at Norfolk, Virginia, in 1933. Unusually, the prototype XP2Y–1 used three engines, the third engine mounted above the main plane. Before long, the squadron had shown the aircraft's worth, making a non-stop flight from Norfolk to Coco Solo in the Panama Canal Zone, covering 2,059 miles in 25 hours, 20 minutes. An even more impressive long-distance flight occurred in 1934, when six P2Y–1s took off from San Francisco on 10th January to fly to Hawaii. Having hit bad weather on the way, which the flight commander, Lieutenant-Commander Kneffler McGinnis, found he could not climb above in his under-powered aircraft, the flight took 24 hours 35 minutes. Altogether, the USN received twenty-four P2Y–1s, with another twenty-four of the uprated P2Y–3, in which the twin inter-wing engines of the P2Y–1 were raised to the level of the upper wing and faired into the leading edge. Exports included the sales of the aircraft to Argentina, Colombia and to Japan, where it was built under licence as the Navy Type 99 or H5Y1 medium flying boat, known later as 'Cherry' to the Allies in World War II.

A further development came with Consolidated's Model 28, an all-metal monoplane flying boat, which the USN initially designated as the XP3Y–1, and this aircraft, as well as having many of the features of the P2Y–1, included a single fin and enclosed positions for the pilots

The Consolidated XP2Y–1 prototype used three engines, reduced to two on production aircraft.

and the gunners. In addition the floats retracted to form wingtips during flight and the engine exhaust was utilized to provide a thermal de-icing system for the first time. The first prototype of this aircraft, using twin 800-hp Pratt and Whitney Twin Wasp radials, was ordered on 28th October, 1933, and built at Buffalo, New York, ready for a first flight on 21st March, 1935.

Although Douglas was working on a competitor to the same USN specification, the XP3D–1, the USN ordered an initial sixty aircraft from Consolidated on 29th June, 1935, re-designating the aircraft the PBY–1. A production PBY–1 cost $90,000 apiece.

Consolidated was more than usually busy at this time. Buffalo was situated in the cold north-east of the United States and could not offer an ambitious flying boat manufacturer open water all the year round, implying the risk of delayed deliveries and broken contracts in a severe winter freeze. In the middle of the XP3Y–1's development, the company moved its factory from Buffalo to the kinder climate of San Diego, California, using a train of fifty boxcars, of which twenty-one included the effects of the employees. The new factory opened on 20th October, 1935. Not all of the employees moved, and one vice-president, Lawrence Bell, stayed behind to open his own aircraft factory in the old Consolidated works. Today, while Consolidated is no more and its successor, Convair, is simply a division of General Dynamics, Bell is one of the world's leading manufacturers of helicopters, along with Sikorsky, of course!

The XPBY–1 used an uprated Twin Wasp engine of 900 hp but, even before this aircraft could fly, the United States Navy ordered fifty of a further development, the PBY–2, in July, 1936. These two versions of the same design were the first of more than 3,000 aircraft which were collectively, in some six versions, to be known as the 'Catalina', a name which some claim originated with the Royal Air Force. While the Catalina, the most numerous flying boat of all time, was primarily a military aircraft, a number were built for commercial customers and for private owners.

An interesting life awaited what would have been the first Catalina to be sold to a civilian customer, an aircraft named *Guba* built for Dr Richard Archbold of the New York Museum of Natural History for an expedition to Dutch New Guinea, which on her delivery flight became the first flying boat to fly across the United States, taking off from San Diego on 24th June, 1937, and arriving in New York a little over seventeen hours later. Before *Guba* could start her expedition, a Russian aviator, Sigismund Levanesky, became lost in an attempt to fly across the North Pole, and the USSR promptly purchased *Guba* to allow Sir Hubert Wilkins, an Englishman, to conduct a search for Levanesky. Wilkins operated *Guba* from a base in the Canadian Northwest Territories, flying 19,000 miles on an unsuccessful search, after which *Guba* passed to her new owners in the Soviet Union. The USSR must have been impressed, for another three aircraft were ordered soon afterwards, being designated GSTs in Soviet Navy service.

A replacement aircraft, also named *Guba*, was purchased by Archbold for the New Guinea expedition, starting on 2nd June, 1938, and making 168 flights in eleven months and carrying almost 300 tons of supplies to field camps, which required the aircraft to achieve such demanding feats as operating from a lake 200 miles inland amidst dense tropical jungle, and, most demanding of all, 11,000 feet above sea level! This aircraft later passed into the hands of the newly formed British Overseas Airways Corporation, BOAC, and then to Saunders–Roe, before finally passing to the Ministry of Aircraft Production, a British wartime Government department, and meeting a sad end, sinking in a storm while lying at anchor in a North Wales harbour.

The uncertainties of the late 1930s ensured a steady stream of repeat orders for the aircraft from the United States Navy. Typical of these was an order for delivery of sixty-six PBY–3s with 1,000-hp Wasps during 1937 and 1938, and for thirty-three PBY–4s with 1,050-hp Wasps, delivered during 1938 and 1939. The last of these was modified as an amphibian, with a tricycle undercarriage, and flew for the first time in this form on 22nd November, 1939, as the XPBY–5A. A standard PBY–5 delivered to the Royal Air Force for evaluation in July, 1939, became the first of many Catalinas to be ferried across the North Atlantic by air. When World War II broke out in Europe in September, 1939, the United States Navy had 224 P2Ys and PBYs of all types in service with nineteen squadrons based at San Diego, Pearl Harbor, Coco Solo, Seattle and Norfolk, Virginia. In the period of little more than two years before the Japanese attack on Pearl Harbor brought the United States into World War II, the USN received 167 PBY–5 flying boats and 33 PBY–5A amphibians. Meanwhile, during the early years of the war, 174 PBY–5s were built for the United Kingdom, France, Canada and Australia, with many of the French aircraft passing to the Royal Air Force on the fall of France in 1940. Most Catalinas were either of the PBY–5 and PBY–5A or the PBY–6 and PBY–6A standard, and the aircraft remained in production throughout the war years, with many being built in the United States and in Canada by Boeing and Vickers to supplement Consolidated's own production, though by the time Catalina production ended, the

The most successful flying boat design was the Consolidated PBY Catalina series; this was a PBY–5A amphibian.

manufacturer had changed its name to Convair. Canadian Vickers-built examples were known as the 'Canso'.

While Catalina production had been building up, other aircraft had also been entering service. A contemporary of the P2Y was the Blackburn Perth flying boat of 1933. This was based on the Iris and retained the biplane configuration with three inter-wing mounted Rolls-Royce Buzzard engines. An all-metal aircraft, or at least one with fabric-covered metal wings and a duralumin hull, the Perth could fly at up to 135 mph and had a range of 500 miles. The anti-aircraft defence of three Lewis guns was complemented by a 37-mm gun in the nose, which could fire a hundred 1½-lb shells per minute, for anti-submarine and anti-shipping operations, although the aircraft could also carry a ton of bombs or depth-charges. It was eventually replaced in 1937 by the Short Singapore, and then by the Short Sunderland.

The United States Navy was not inclined to limit itself to the PBY Catalina series, no matter how good, and indeed the service was actively looking for a large four-engined flying boat for patrol duties, evaluating rival Sikorsky and Consolidated designs in this task. Sikorsky produced the prototype XPBS–4, using four 1,050-hp Pratt and Whitney Wasp radials, which flew for the first time on 13th August, 1937. A high-wing monoplane with a single-fin tail, this was

The larger Consolidated, or Convair, PB2Y–4 Coronado, originally developed as a four-engined Catalina, 'on the step' before take-off.

the first aircraft in the United States to be fitted with 0.50-calibre machine-guns, and was known as the 'Flying Dreadnought', although it was also reasonably fast for such a large aircraft, with a maximum speed of 230 mph. The prototype rival Consolidated XPB2Y–1 was effectively a four-engined Catalina, first-flying on 17th December, 1937, but a twin-fin was added later to provide improved directional stability. This was another contest won by Consolidated, but the victory was somewhat academic, for although six PB2Y–2s were ordered at a cost of $300,000 apiece, it was decided that larger numbers of Catalinas, at a third of the price, were a better alternative. The PB2Y–2s were fated to act as development prototypes for the XPB2Y–3s, using four 1,200-hp Pratt and Whitney Wasps and with a range of 2,300 miles. Known as the 'Coronado' to the Royal Air Force, deliveries of the production PB2Y–3 started in 1942, with 170 being built for the USN and the RAF. The RAF only received ten out of an order for thirty-three, and although the aircraft was intended for long-range maritime reconnaissance carrying up to four 325-lb bombs or depth-charges, the fate of most of the RAF aircraft was to be used as transports on the long Pacific routes.

Consolidated had hoped for a USN or a Pan American contract for a large flying boat based on the aerodynamic surfaces of the B–24 Liberator bomber which, in civil use, would have rivalled the Boeing 314 and which, in its naval version, would have competed with the new Martin design, the XPBM–1 Mariner. However, the demands of Liberator production prevented this aircraft from being put into production.

The Martin XPBM–1 was a large twin-engined flying boat with a gull-wing and a twin-fin tailplane. It was one of the few aircraft to have had a scale model built, a small quarter-scale prototype which flew, although, being a small aircraft with the proportions of a large flying boat, it looked uncanny. In common with the Sikorsky XPBS–4, the Mariner used 0.50 calibre machine-guns in its nose and tail turrets, as well as in two waist gun positions. First flown in full-sized form on 18th February, 1939, the aircraft could lift 3½ tons of armaments and fly at ranges of up to 3,000 miles with a 200 mph maximum speed. Although early versions had retractable floats, most used fixed wingtip floats. The aircraft was powered by two large 1,600-hp Wright Cyclones, and was built in quantity, with some six hundred built by the end of 1945. Later versions were equipped with radar and some of the other anti-submarine warfare aids developed during World War II, often with the anti-submarine search radar in a 'conning-tower' rising above the fuselage amidships. In spite of the large number built, the aircraft was unpopular with aircrew, being known as the 'flying gas tank', due to the tendency to accumulate gasoline fumes inside the fuselage – which ensured that there was a strict 'no smoking' ban for aircrew, and frequent reminder notices aboard! Only the USN operated the Mariner. The small number delivered to the RAF never entered service. Later versions included the PBM–5A amphibians.

Another successful World War II flying boat was a development of Short's S.23 Empire 'C' class flying boats built for Imperial Airways and Qantas, the S.25 Sunderland, which on entering service with the RAF was to be that service's largest aircraft, and even today has the distinction of being the longest serving aircraft type in RAF service. A large four-engined long-range maritime reconnaissance flying boat, the Sunderland incorporated many years of experience in its design and was renowned for its strength, a necessity for an aircraft operating for much of its time over the stormy reaches of the North Atlantic. The Sunderland was well equipped with a strong defensive armament, with machine-guns in the nose, amidships and tail positions, earning it the nickname *Stachelschwein*, or 'porcupine', from Luftwaffe pilots.

The prototype Sunderland first flew on 16th October, 1937, powered by four 950-hp Bristol Pegasus radial engines, but the power output was increased on the early production versions which used 1,010-hp Pegasus engines. The early armament of four 0.303 Browning machine-guns in the tail, a single 0.303 Vickers machine-gun in the nose, and two more Vickers machine-guns in the mid-upper position, was later varied, usually with four machine-guns in the nose and tail turrets, and two in the mid-upper turret. Capable of carrying

The Martin PBM–1 Mariner was one of the first flying boats to be equipped with radar and was also available as an amphibian, the PBM–5A.

up to 2,000 lbs of bombs or depth-charges, the Sunderland had a cruising speed in excess of 200 mph and a range of 2,400 miles, while later versions were fitted with radar, although, of course, this still limited aircraft to finding submarines on the surface. Deliveries of the Sunderland started in 1938, with the first aircraft joining No.230 squadron, RAF, in Singapore. By the outbreak of war in Europe, the RAF had forty Sunderlands in service. Wartime production stretched Short's resources to the limit – the company was also building the Stirling bomber – and in spite of the opening of a new factory at Belfast in Northern Ireland, owned jointly with the shipbuilders, Harland and Wolff, the company becoming known as Short Brothers and Harland. The Sunderland was also built in England by Blackburn before and during the war years.

Anti-shipping radar was introduced with the Sunderland II in 1941, while the Sunderland III introduced a redesigned hull with improved planing capabilities. The Sunderland IV used 1,700-hp Bristol Hercules engines, although on entering service this aircraft was renamed the Seaford. Post-war, the aircraft was the progenitor of many civil conversions and developments for BOAC and other airlines, while the Royal Australian Air Force, Royal New Zealand Air Force and the Aéronavale were also enthusiastic operators of the Sunderland. Sunderlands remained in RAF service until 1958, with the Aéronavale until 1960, and the RNZAF until 1966. Some six hundred aircraft were built between 1938 and 1945.

The long-range maritime-reconnaissance flying boat of World War II was far superior to its World War I counterpart in every respect, almost to the extent of being counter-productive as aircrew on the long ocean patrols sometimes would hallucinate during the tedium of hours spent flying over a featureless ocean. However, just as the flying boat reached the peak of its performance it was threatened by the rapidly improving performance of landplanes, such as the Junkers Ju.290 and the Focke-Wolf Fw.200, or by maritime reconnaissance developments of landplane bombers.

Developed from the Short Empire series, the Short Sunderland had a distinguished record during World War II and for many years afterwards.

7
WORLD WAR II

Even at the onset of hostilities when World War II broke out in Europe during September, 1939, the flying boat was a highly sophisticated and vital element in the maritime defences of the leading combatants. It was an invaluable search and rescue aircraft, both in its pure flying boat form and as an amphibian, or even as a seaplane, and its applications for transport, communications and reconnaissance should not be overlooked. Significant also for the seaplane was the fact that most major warships still lacked radar, and while at first this new invention was confined to British warships, it had limitations of range, which meant that the seaplane could still undertake longer-range reconnaissance.

The growing capabilities of landplanes and their greater speed already showed that the position of the flying boat was no longer secure, but in a fast-moving war the ability of the flying boat to free air forces and navies from the need to operate only from bases with concrete runways remained an advantage throughout the war years. The seaplane, however, had already lost its position in the fighter, bomber and torpedo-strike squadrons due to the greater superiority of the carrier-borne landplane and the ability of the more modern aircraft-carriers to operate comfortably with the rest of the fleet in a way that the seaplane-carrier generally found difficult. Such were the advances in naval aviation that much of World War II was a war between carrier fleets rather than between battle fleets.

Aircraft such as the Catalina and the Sunderland were already well in advance of the biplanes of the still recent past in USN and RAF service, and indeed of the early war years for other air arms, including the Royal Canadian Air Force. The toughness of these two aircraft provided welcome extra security for crews forced to land in the open sea, either to rescue the survivors from hostilities – perhaps from a ship which had been sunk or an aircraft shot down – or even because the flying boat itself had been shot down or seriously damaged, or had suffered mechanical failure. Mechanical reliability was a significant improvement over that of earlier aircraft but still far below the standards taken for granted today.

For the early combatants, the war was by no means a surprise. The Munich Agreement of 1938 has been much criticized, especially for its naïvety, although one could argue that it was an honest mistake of democratic statesmen forced to negotiate with a dictator with all the power of a twentieth-century military machine behind him. However, there can be no doubt that Munich gave both sides the extra time that they wanted, with the Germans consolidating their position in central Europe and building up their armed forces to a new peak, and the British doing much to repair the damage caused by disarmament policies between the wars and of course by reductions in spending during the Depression years. Only the French gained little benefit from the respite gained at Munich, the decline of the French defensive forces having been taken too far. There was also an undue and ill-deserved faith in the defensive qualities of the Maginot Line, and the aircraft industry, which had been nationalized, was in a state of chaos which prevented it from reaching the peaks of production gained by its British counterpart. However, for the most part, the aircraft in service in France were simply improved pre-war designs.

The Germans had to some extent neglected their maritime reconnaissance squadrons during the late 1930s, with the emphasis of reconstruction in the Luftwaffe being placed upon strategic and army support aircraft, while the German Navy, shorn of its own air power, could do little to press the case for a stronger long-range

A Luftwaffe Blohm und Voss Bv. 138 flying boat ready on the catapult.

reconnaissance and anti-shipping air arm. However, there was a belated modernization of the maritime reconnaissance elements during the last year or so of peace, and indeed at sea the Arado Ar.196 was replacing the few remaining elderly Heinkel He.60 seaplanes while ashore the He.59 had finally given way to the new He.115. Blohm und Voss Bv.138s and Dornier Do.18 flying boats were in service, joined by requisitioned Deutsche Luft Hansa Do.26 flying boats for troop carrying and other transport duties.

In France, the piecemeal procurement policies of French governments between the wars had meant that there were relatively few aircraft in Aéronavale service and, for the numbers, there were too many types. Older aircraft included the Breguet Br.521 Bizerte flying boat as well as the Latécoère 302 and slightly older Lioré 70 flying boats. These were soon joined by small numbers of Potez-CAMS 141 and Sud-Est LeO H–246 four-engined flying boats and by the six-engined Latécoère 523, as well as by Latécoère 298 twin-float seaplanes, which were operated from the seaplane-carrier *Commandant Teste*, commissioned during the hey-day of the seaplane, in 1929.

While the Royal Air Force also had some older designs in service, it was well advanced in its re-equipment programme as deliveries of the Short Sunderland built up, and deliveries of the Consolidated Catalina

Breguet Br.521 Bizerte flying boats were an important part of the Aéronavale strength at the outbreak of war.

Another picture of a Br.521 Bizerte, clearly showing its Calcutta ancestry.

were soon due, especially after May, 1941. The RAF was also able to take advantage of certain other factors to build up its maritime reconnaissance effort as war began. As an example, the first Royal Australian Air Force Short Sunderland squadron was formed in Britain at the onset of war, and the Australian Government promptly transferred this unit to the RAF, complete with aircraft and personnel! Similarly, on delivery during late 1939 and 1940, the Imperial Airways 'G' class flying boats were transferred to the RAF for use as VIP transports, and indeed a number of the 'C' class Empire flying boats were also handed over to the Royal Air Force, while other BOAC aircraft were operated by their civilian crews on RAF transport flights. Both the RAF and the Royal Navy took delivery of Supermarine Walrus amphibians for search and rescue, while the Royal Navy also received a few float-equipped Fairey Swordfish biplanes.

At this stage, the United States, Japan and Italy were not involved in World War II, although again the signs of future conflict were there, and these countries showed strong preferences amongst the warring nations of Europe. The United States was disposed to take the side of Britain and France, fellow democracies, while Italy favoured the style and the policies, if in a less brutal form, of the German leadership. Alongside this, both Italy and Japan wished to pursue colonial policies in North Africa and China respectively.

Too late, the Breguet Br.730 was an up-to-date flying boat in production at the time of the fall of France.

In Italy, the next country to enter the war, the Regia Aeronautica was well supplied with the CRDA Cant.Z.501 Gabbiano flying boat and the Z.506B Airone seaplane, which were complemented by a number of the older S.M.55 flying boats, and airline S.M.66s were soon requisitioned for wartime search and rescue and transport duties.

At first, there was what the British aptly described as the 'phoney war', with relatively little action, but during 1940 the tempo of the conflict was dramatically accelerated after the Germans completed their conquest and occupation of Poland and turned their attention to the west. The first major advance came on 9th April, 1940, when German forces attacked Denmark and Norway simultaneously. The poor state of Denmark's defence meant that the country quickly succumbed to the fast German advance, but in Norway, with the help of a British Army Expeditionary Force and inhospitable terrain, there was a stiff and bloody resistance to the German invasion. Relatively few British flying boats or seaplanes took part in the Norwegian campaign, largely because the Royal Navy was able to deploy three of its aircraft-carriers off Norway, although the Fleet Air Arm's obsolescent aircraft were easy victims for the well-equipped Luftwaffe squadrons. Some thirty Luftwaffe Heinkel He.115s and a number of transport Dornier Do.26 flying boats were part of the overall German deployment of some four hundred aircraft over Norway, facing the

A development of the Breguet Br.730 was the Br.731 Vega, which fell into German hands, although this example looks as if it may have been 'liberated'.

Royal Norwegian Air Force's small fleet of four Heinkel He.115s and a number of older aircraft, though two Luftwaffe He.115s and an Arado Ar.196 from the cruiser *Admiral Hipper* were captured by Norwegian forces and pressed into service. The Luftwaffe also had a small number of Blohm und Voss Bv.138s, especially equipped with degaussing equipment to explode magnetic mines.

An attack on Belgium and the Netherlands followed, while France was also attacked, mainly through Belgium, from 10th May onwards, with Aéronavale units redeployed to army support duties in a frantic attempt to hold back the fast German advance. The entry of Italy into the war on 10th June meant that the Aéronavale had also to make aircraft available for patrols in the Mediterranean against the Italian Navy. After the French surrender on 22nd June, the British forces were forced to retreat from Dunkirk, and a number of French aircraft were flown by their pilots to Britain to continue the war, although others fell into German hands, and others still were absorbed into the forces of Vichy France – a substantial number of which were stationed in North Africa. One of the more tragic episodes of the war followed the fall of France, as the British sought to prevent the powerful French fleet, most of which was in North Africa, from falling into German hands. French warships were sunk at Mers el Kebir and Dakar. Although flying boats were not involved in this action, a Lioré 130 seaplane from the battleship *Strasbourg* was shot down by a Fairey Swordfish after it had tried to attack the aircraft-carrier, HMS *Ark*

One of the most impressive World War II flying boats was the Blohm und Voss Bv.222 Wiking, a large six-engined German transport flying boat.

Royal. In another incident, two Br.521 Bizerte flying boats were shot down by British fighters after shadowing the British fleet off Gibraltar.

After the outbreak of war and before the fall of France, several new French flying boat types had entered service, including the four-engined Breguet Br.730 and its development, the Br.731, and the Latécoère 611, but these were all too few and too late, and with many other French aircraft, they were destined to fall into German hands. However, a Latécoère 302, with a single bow gun and twin amidships positions, fled to Dakar, and once there maintained regular patrols over the Mediterranean until being scrapped in late 1941 due to a shortage of spares.

The Germans introduced a number of significant new aircraft types during the war years, but relatively few of these were flying boats or seaplanes, and indeed the only hydro-aeroplane of any real importance was the giant Blohm und Voss Bv.222 Wiking, or Viking. Originally designed to be the German contender in the race to build a non-stop transatlantic passenger-carrying flying boat, the Wiking was, of course, intended to be operated by Deutsche Luft Hansa. However, the Bv.222 never entered service with DLH, not making its first flight until well after the outbreak of war, on 7th September, 1940, and then being delivered to the Luftwaffe as a military transport in July, 1941. The prototype was unarmed, as befitted an aircraft originally conceived as a commercial flying boat, but later aircraft were equipped with MG.131 machine-guns in the nose, tail and mid-upper positions.

A total of just thirteen Bv.222s were built, and indeed the aircraft was really too big for effective use. It must have been hard to justify the resources required to maintain even the limited production which was achieved when many smaller aircraft could have been built instead for a more immediate and constant benefit to the German war effort. Six 1,000-hp BMW Bramo Fafnir radial engines were used in this large monoplane, which nevertheless had a maximum speed of 220 mph and a range of some 4,500 miles, as was necessary for direct flights from Germany across the Atlantic. In spite of a wingspan of 151 feet and a length of 120 feet, the aircraft, as originally conceived for DLH service, was only intended to carry twenty day or sixteen night passengers. The Wikings were used for a short time as long-range maritime reconnaissance aircraft during 1943, under the command of Fliegerführer Atlantik. A more powerful version to the Wiking, the Bv.238, did not fly until 1944, and only the prototype was completed – two more aircraft were uncompleted at the end of the war – but was destroyed when a formation of USAAF North American F-51 Mustang fighters discovered the aircraft between test-flights at its lakeside base. Only three of the standard Wikings survived the war, two passing to the United States and one to the United Kingdom, for post-war examination and experiment.

Less massive than the Bv.222 Wiking, but far more novel, was the British Blackburn B.20, a flying boat with the unique feature of a retractable hull. Designed to overcome the major problem of the flying boat, the drag of the deep hull once airborne, in flight the B.20 looked like a conventional bomber, while on the water it looked as if the bottom of the hull and the main fuselage had been separated by struts. Apparently handling well, both on the water, with the retractable hull

The Blackburn B.20 was unique: the hull retracted to provide a more streamlined shape while in flight, yet still keeping the propellers clear of the water while taking off and landing.

For some time after the outbreak of war, the Royal Canadian Air Force continued to operate Supermarine Stranraer biplane flying boats: this one is visiting the US Navy at Pensacola.

keeping the propellers clear of the water, and in the air, the B.20 was a medium-sized aircraft for its day, with a wingspan of 82 feet and a fuselage length of 70 feet. Two Rolls-Royce Vulture engines gave it the high speed for a flying boat of 322 mph and a range of 1,500 miles. Jacques Schneider would have been delighted with such an aircraft, which was a real competitor in performance for many landborne bombers of the day, but more urgent priorities and simpler and less expensive aircraft of a more conventional configuration meant that only the prototype B.20 was ever built. Eventually, some time after test-flying started in 1940, the prototype itself was lost on a test-flight.

Meanwhile, German forces, having secured their position in the west, had pushed south, and although an earlier Italian attempt at invading Greece had been repulsed by British and Greek forces, the Germans themselves succeeded in occupying Greece in April, 1941. While there were relatively few opportunities for hydro-aeroplanes to be deployed in this theatre of war, the Royal Hellenic Air Force was still operating a small number of elderly Fairey IIIF seaplanes, as well as some more modern Dornier Do.22 flying boats. In June, 1940, an end to the alliance between the Soviet Union and Germany came with Operation Barbarossa, with German forces pushing into Soviet-occupied Poland and then into the USSR itself.

Although the United States did not enter World War II until the end of 1941, the United States Navy, and indeed the rest of the American armed forces, were allowed to expand and modernization was accorded a high priority as soon as war broke out in Europe. The situation was far from being a repeat of World War I, even in Europe, and the growing threat to peace and stability in the Pacific from Japanese expansion was also recognized by the Americans. The threat to

Vought OS2U seaplanes flying in formation – this was the standard USN shipboard reconnaissance seaplane at the outbreak of World War II.

Britain, the anger over German treatment of Jewish communities, and a general lack of sympathy with German aims, all contributed to a situation in which the United States could have been dragged into the war, even without the Japanese problem.

Production of Consolidated Catalina and Martin Mariner flying boats forged ahead, with Boeing providing additional production of the Catalina. Vought OS2U Kingfisher observation seaplanes were also being built for operation from warships, although many of the new Cleveland class of cruisers were soon modified, most becoming the Independence class of light fleet, or CVL, aircraft-carriers. In February, 1941, Congress approved the Lend-Lease Act, allowing the United States to provide military equipment for Britain on favourable terms, and of course the Catalinas for the Royal Air Force were one of the more prominent items covered by this generous scheme. The Japanese had been spending heavily on armaments throughout the period from the mid-1930s onwards, pressing ahead with a war of conquest in mainland China, despite American objections. Ultimately, the Americans were forced to impose economic sanctions against Japan.

Unlike the change in American, British and European practice by this time, a number of Japanese fighter and bomber aircraft were available in either landplane or seaplane form, but for operations from battleships, cruisers and the seaplane-carriers *Chitose* and *Chiyoda*, the Japanese used Aichi E13A1 Type O three-seat reconnaissance seaplanes. Meanwhile, ashore they used Kawanishi H6K2 Type 97 and H8K1 Type 2 flying boats, the latter being given the codename 'Emily' by the Allies.

Before the outbreak of war in the Pacific, the United States Navy was further reinforced in November, 1941, when, by presidential decree, the United States Coast Guard was transferred to USN control, a normal preliminary in wartime. This provided the USN with additional aircraft, including Hall PH–2 and PH–3, Viking OO–1 and General PJ–1 flying boats, and Douglas RD–4, Grumman JF–2, JRF–2 and J4F–1 amphibians, as well as a few Catalinas, although all of these aircraft were available only in small numbers, and some were fairly elderly. The absorption of the USCG into the USN provided an integrated control over the sea around the United States and, of course, furnished the USN with aircraft for coastal patrols, while the men and equipment provided, with ships as well as aircraft, could be deployed as a single force.

Japan's most modern flying boat at the outbreak of war in the Pacific was the Kawanishi H8K1 Type 2 'Emily', first flown during 1941, although at first there were problems with stability on the water and a tendency to porpoise. Modifications followed, and these were incorporated on the production version, the H8K1 Model 11, which had a deeper hull and improvements to the keel, as well as knuckles, to ensure better handling during take-off and landing. This aircraft joined Japanese Navy Air Force squadrons later in the year. The aircraft used four 1,530-hp Mitsubishi Kasei radial engines. Only a few of the H8K1 version of this aircraft were built, but during the war years more than a hundred of the improved H8K2 which followed were built, and there were thirty-six H8K2–L sixty-four seat transport flying boats as well, known as the 'Sei-Ku' or 'Clear Sky'. Two further developments, the H8K3 and H8K4, never entered production. The

Backbone of the Japanese Navy's maritime-reconnaissance element in World War II was the Kawanishi H8K2 Type 2 'Emily' flying boat.

H8K2 used uprated 1,850-hp Mitsubishi Kasei radials, enabling the aircraft to lift 4,000 lbs of bombs, mines or depth-charges. A formidable defensive armament was provided, with twin 20-mm guns in the nose, tail and mid-upper turrets, while there were other gun positions on the flight deck and amidships, with 7.7-mm machine-guns. Before the end of the war, some of these flying boats were equipped with anti-shipping radar.

The backbone of the Japanese Navy's flying boat strength at the beginning of the war in the Pacific was the Kawanishi H6K5 'Mavis', with a span of 131 feet and a maximum speed of 230 mph, and which in general had an appearance not unlike that of a four-engined Catalina.

A far smaller flying boat than the 'Emily' was the small Grumman G–44 Widgeon, in fact an amphibian developed from the Grumman G–21 Goose, and a smaller version of that aircraft. First flown in July, 1940, the twin-engine G–44 was initially intended as a four- or five-seat private-owner and light-commercial amphibian, but the United States Navy ordered twenty-five as the J4F–1 for search and rescue operations, with just two or three seats. The USN also experimented with the Widgeon for anti-submarine duties, but the aircraft was really too small to be of much use for this type of work, even though one did sink a U-boat with a 200-lb depth-charge in 1942. Most USN J4F–1s eventually operated in a training role, while the USCG used the aircraft for coastal search and rescue. The USAAF received ten J4F–1s, which they redesignated the OA–14, and the RAF also received a small number of these aircraft, which in British service became known as the Gosling. A redesigned hull led to the post-war G–44A Super Widgeon, built by Grumman and in France by SCAN, using two uprated Lycoming engines.

An 'Emily' in trouble flies low to avoid attacking aircraft.

The other major Japanese maritime reconnaissance aircraft was the Kawanishi H6K5 'Mavis', similar in appearance to a four-engined Catalina.

All of the US armed forces used the small Grumman J4F–1 Widgeon during World War II – usually for training. This is a USAAF example.

Naturally, the United States Navy was also interested in larger flying boats than the Widgeon, and one large and long-range maritime reconnaissance flying boat tested during the war years was the twin-engined Boeing XPBB–1 Sea Ranger. Although the first flight during summer, 1942, was a success, this aircraft, with a wingspan of 139 feet and eight defensive gun positions, was not put into production because of the effect which it would have had on production of the USAAF's Boeing B–29 Superfortress heavy bombers.

War had come to the Pacific and indeed the war became a 'world war' on 7th December, 1941, when some 350 aircraft from six Japanese aircraft-carriers attacked the American naval base at Pearl Harbor, on the island of Oahu in Hawaii. The declaration of war on the United States by Japan followed the attack! The attack came as a surprise, although warning signs were there for those alert enough to notice and to take action, with a number of incidents on the day of the attack being overlooked or misinterpreted. Not the least of these was the detection of a Japanese submarine off the Hawaiian islands on the day of the Japanese attack, but the crew of the Catalina flying boat which discovered the submarine decided that it must have been a USN vessel because of its proximity to the naval base! The mistake, and others on the same day, proved to be costly – out of thirty-six Catalinas based at Pearl Harbor, only three survived the attack.

The Japanese swept through the Pacific region during 1942, with Malaya and Singapore falling in February, followed by Burma in March, and then Sumatra and Java were invaded. Only in the Philippines was serious resistance possible, before they too were overrun. A number of Dutch Catalinas and Dornier Do.24s operated by the Netherlands East Indies Naval Air Service were evacuated to Australia as Sumatra fell, and these aircraft with their crews were redeployed to Ceylon before eventually rejoining Dutch groups in the United States and the United Kingdom. Under pressure on all fronts, the Allies also had to protect vital convoys in the Atlantic and the Mediterranean, and to a lesser extent in the Pacific – in the latter theatre of war it was at least possible to avoid Japanese aircraft, while the Japanese were never as effective in using the submarine against convoys as the Germans proved to be in the North Atlantic. Increasingly, many of the maritime reconnaissance aircraft and the anti-shipping aircraft of World War II were large landplanes, such as the specialized German Focke-Wulf Fw.200 and the Junkers Ju.290. These were developed from airliners much in the manner of modern maritime reconnaissance aircraft such as the Lockheed P–3 Orion, developed from the Electra, and the British Aerospace Nimrod, developed from the de Havilland Comet airliner. The USN found itself making greater use of the Consolidated Liberator and the RAF in a similar position was using Vickers Wellingtons, supplementing the efforts of the flying boats, although of course the implications for the future of maritime reconnaissance were there to be seen.

The flying boat still bore the brunt of the long patrols over the Atlantic and the Pacific. One of the early notable achievements of the Catalina in RAF service came in May, 1941, when aircraft based at Gibraltar helped in the destruction of the German battleship *Bismarck*. Apparently, the unit had a number of USN officers helping in the conversion training of the RAF personnel, and one American officer found himself on a patrol which discovered the German ship's position. At the outbreak of war in 1939, the RAF had three Sunderland squadrons, Nos. 203, 210 and 221, and an RAAF squadron became a fourth Sunderland squadron, No. 10. By the end of the war, there were thirty Sunderland squadrons in RAF service – the flying boat may have been under challenge, but it was far from being outclassed completely at this stage!

The first operational flying boat patrol of World War II was operated by a Royal Air Force Short Sunderland of No. 210 Squadron on 3rd September, 1939, and the following day another RAF

Sunderland became the first British aircraft to be attacked by anti-aircraft fire, although on this occasion it was British AA fire from gunners who had mistaken the aircraft for a German machine. Fortunately, they did little damage. A few days later, on 8th September, another 210 squadron flying boat located and attacked a German U-boat, but failed to sink it. A humanitarian act followed on 18th September, when two RAF flying boats from 228 squadron landed in rough seas to rescue thirty-four survivors from the British merchantman, *Kensington Court*, which had been sunk by a U-boat. Later, after 228 squadron had received its Sunderlands, it was one of their aircraft which scored the first near-kill of a U-boat, surprising U-55 after it had been forced to surface by Royal Navy escort vessels. The arrival of the Sunderland finally persuaded the U-boat's commander that it would be better to scuttle his craft rather than face further attack. It was not in fact until 17th July, 1940, that one of the Australian Sunderlands of No. 10 Squadron sank the Sunderland's first U-boat. One of the problems of hunting small submarines in a wide ocean with still limited detection equipment can be judged from the following comment by the aircraft's captain in a BBC radio interview: 'I have heard of crews that go out for a year and have not even seen a U-boat, so we have been rather lucky.' The crew had done well, luck or not, because the port bomb racks were defective and, under intense fire from the U-boat, they had had to transfer bombs by

A formation of Royal Australian Air Force Consolidated Catalinas.

hand from the port racks to the empty starboard bomb racks.

Life in a long-range maritime reconnaissance aircraft could be boring, but it was also vitally important. At worst, it forced U-boat commanders to keep their heads down; at best, it was another potent weapon able to help counter a serious menace to Allied shipping which, if it had gone unchecked, could have lost Britain the war. The shortcomings were those of the technology then available, but surface vessels also suffered from the limitations of their equipment: there were no magnetic anomaly detectors or sensors for diesel fumes, let alone sonobuoy systems such as 'Jezebel', all of which would have made detection of diesel-electric submarines so much easier. As it was, the angular superstructure of the World War II submarine would have provided plenty of signals for modern detection equipment, but the surface vessels lacked the anti-submarine helicopters and variable depth sonar of today, and neither aircraft nor ships could deploy homing torpedoes against submarines.

Long and cold flights in a flying boat were only for those with the physical toughness to overcome the ordeal and the mental strength to cope with unremitting tedium without becoming lax and inefficient. When something did go wrong, however, the Sunderland could prove that it was a strong aeroplane, well suited to the worst the North Atlantic could offer. In October, 1940, a Sunderland was two hours out from its base on a patrol when a sudden electrical storm destroyed the compass and seriously damaged the radio. The crew could not send signals for another five hours, and even then could not receive them. A further five hours flying and they were completely lost, but had to land before their fuel ran out, since the Sunderland did not 'glide' too well, and landing without power in an 80-mph gale and with 20-foot-high waves would have been even more difficult, if not completely impossible, to achieve safely. As soon as the aircraft landed, the crew were prostrated by violent seasickness, which they suffered for several hours, but the wireless operator managed to continue sending signals without at first knowing whether or not they were being received, although later they picked up a message advising them that a warship could arrive in eight hours, while they continued to drift helplessly before the fury of the storm. As luck would have it, the warship arrived as the aircraft started to break up, flinging her crew into the sea. It was another fifty minutes before the warship was finally in a position to rescue the Sunderland crew, sending a commander and twelve ratings over the side with ropes tied to them because it was impossible to launch a boat in conditions in which waves were reaching bridge height as they swept over the ship, the cruiser HMAS *Australia*. They

Launch for a Short Sunderland – the forward gun turret has been recessed so that the aircraft can be moored.

EJ164

succeeded in rescuing nine out of the thirteen men aboard the Sunderland, who had spent nine hours aboard the aircraft and almost an hour in the sea, after twelve hours in the air!

For the first few years of the war, the Sunderland was the only aircraft in RAF Coastal Command capable of extended long-range maritime reconnaissance, operating over a radius in excess of 500 miles, although the addition of the Catalina helped to improve the situation to some extent. Even though the United States Navy had based patrol aircraft in Iceland before their entry into World War II and, when they did enter, along with Canada provided aircraft patrols from the western side of the Atlantic, convoys were still largely unprotected from the air for much of the distance across the North Atlantic. Before deliveries of the Catalina to the RAF began to build up, a Saunders–Roe, or SARO, flying boat was evaluated for maritime reconnaissance, but this aircraft proved to be unsuitable, and was not ordered into production.

In the Pacific, before the outbreak of war, the United States Navy had had aircraft not only in Hawaii but also in the Philippines, where Patwing (patrol wing) 10 had two squadrons, each with fourteen PBY Catalinas. These aircraft suffered heavy losses in the Japanese attack and subsequent invasion of the Philippines. Immediately before the American evacuation of the islands, just three aircraft were left to take senior USN and United States Army officers off the islands, but the remaining Catalinas were themselves heavily damaged in a bombing raid, leaving evacuation of senior personnel to be conducted by submarine. But the tide of war turned more quickly in the Pacific than elsewhere, at the Battle of Midway. It was a USN Catalina which first spotted the Japanese fleet on its way to Midway, leading to a resounding Japanese defeat on 4th June, 1942. As the war zone spread, USN aircraft were also based in Australia, with Patwing 10 moved to Perth in Western Australia, from whence patrols could be made over the Indian Ocean, and Patwing 17 based at Brisbane, for Pacific and Tasman Sea patrols. Radar-equipped 'Black Cats', Catalinas painted black, were used by VP–12 and VP–54 to patrol the sea between New Georgia and Santa Isabel through which the Japanese were running supplies to Guadalcanal. Another new station for the USN was Brazil, with Martin PBM Mariners of VP–83, 74 and 94 based in that country for patrols over the South Atlantic.

The Catalina was a much improved aeroplane by the end of the war. The list of modifications to what was already an excellent aircraft included the provision of limited self-sealing tanks for part of the fuel load. Patrol squadron VP–63 was the first to operate aircraft with

A Sunderland planes 'on the step' before 'unsticking'.

magnetic anomaly detectors on its PBY–5A amphibians, and was also the first USN squadron to operate from the British Isles, starting on 23rd July, 1943, to maintain patrols over the Bay of Biscay. In January, 1944, a squadron of Catalinas was based in Morocco, this was VP–63 again, to patrol the Straits of Gibraltar. It succeeded in sinking three U-boats during 1944, while Catalinas accounted for fourteen U-boats in the North Atlantic between August, 1942, and July, 1943. As demand built up, Boeing and Vickers in Canada built the Catalina, with the first Canadian-built Boeing PBY–5A, known as Canso A, going to the RCAF and flying on 27th July, 1942. Vickers' Canadian factory at Montreal became known as Canadair after December, 1944. A number of Catalinas were also built at the Consolidated Vultee factory in New Orleans, and the Naval Aircraft Factory built a small number, initially seventeen aircraft, as the PBN–1 Nomad. The tailplane developed for the machine was adopted for the PBY–6A amphibian, the definitive aircraft in the Catalina series.

Even the United States Army Air Force operated a small number of Catalinas, mainly on search and rescue operations in the Pacific, designating the aircraft as the OA–10, and when the PBY–6A entered USAAF service it was designated OA–10B. In Royal Air Force service, eventually some six hundred Catalinas were operated in twenty-one squadrons, based round the world from Murmansk in Russia to Singapore.

A total of 2,160 Catalinas were built at the San Diego factory of Consolidated, or Convair, 731 by Vickers or Canadair and Boeing in Canada, 235 by Consolidated–Vultee in New Orleans, and 155 by the Naval Aircraft Factory at Philadelphia. The Soviet Union was given almost two hundred Catalinas during the war, while smaller numbers of aircraft entered service with other Allied nations, including, for example, thirty-four aircraft delivered to the Royal New Zealand Air Force.

There were surprisingly few significant operations by Japanese flying boats during the war. However, one is certainly worthy of mention: the attack by three 'Emily' flying boats on Pearl Harbor on 4th March, 1943, after the aircraft had refuelled from submarines with which they had made a rendezvous at French Frigate Shoals. A similar attempt about a year earlier had been defeated by bad weather. Though an impressive feat, the strategic value of such a raid was obviously limited, serving only to surprise and lessen the confidence of the American personnel at Pearl Harbor, and obviously the damage which could be inflicted was relatively slight.

Naturally, the outbreak of war had brought an end to the rapid pre-war expansion of flying boat services by the British Overseas Airways Corporation, or BOAC, Pan American Airways, Air France and Deutsche Luft Hansa. Even neutral countries found air services severely restricted by hostilities, and of course by the shortages of fuel and spares which also resulted from a war in which the major aircraft-manufacturing nations were fully absorbed. But this did not entail the scrapping of commercial flying boats; they were soon put to a variety of different uses. The ex-DLH Dornier Do.26s and Blohm und Voss Bv.138 flying boats were soon operating as military transport aircraft in the invasion of Norway, while other Bv.138s were used for mine counter-measure duties. The Italians also used converted civil flying boats for transport and search and rescue, and for maritime reconnaissance in the Mediterranean and the Adriatic. As a result, the Savoia–Marchetti S.M.66 was pressed into service and the Z.501 flying boats and Z.506 seaplanes were freed for anti-submarine and anti-shipping operations.

While many flying boats were requisitioned and operated by military or naval personnel, others were operated for the military by their civilian crews. In other instances, airlines continued to operate on special flights maintained by governments for diplomatic and strategic reasons, and of course on which normal peacetime civilian passengers could not be carried. The importance of air transport was well understood, and, in addition to its speed, there was the relative freedom from enemy attack, on the basis that attack by fighter aircraft on a flying boat was less of a hazard than submarine attack on a merchant vessel; the latter could happen almost anywhere, while fighter aircraft were restricted by the bases available, except in the Pacific where, alone amongst the Axis powers, the Japanese had aircraft-carriers.

Even before the attack on Pearl Harbor, but in that same fatal month, the United States Navy had formed a Naval Air Transport Service (NATS), and this was ready for operations by the following February, using Catalinas and PBM–3 transports as well as several landplane types. Indeed, during the war in the Pacific, the flying boats proved their worth, making operations possible from many small islands which lacked runways or where runways had been put out of action by enemy bombing, or where the runways were too short for heavy transport or maritime reconnaissance aircraft. The Martin Mars was especially suited for transport operations in the Pacific, and many of these aircraft joined NATS as the war progressed.

A large twin-finned four-engined flying boat, the Martin Mars, was developed as the Martin Model 170 long-range maritime reconnaissance flying boat, first flown in late 1941, with production aircraft designated the XPB2M–1 Mars. Four 2,200-hp Wright Double Cyclone radial engines, a wingspan of 200 feet, a length of 117 feet, and a maximum speed of 210 mph gave it a range of 4,400 miles. In common with most larger flying boats, the Mars had two decks, but the aircraft was designed with only a limited defensive armament and little armour protection. Having been designed in this way, it was decided that there was neither the time nor sufficient need to redesign the aircraft to meet combat requirements, and instead it was put into production as a long-range transport aircraft, in effect making a virtue out of necessity, although the availability of Catalinas and Mariners in large numbers obviously made the decision that much easier. The Mars transport aircraft were operational by December, 1943, and on the first naval cargo mission, a Mars flew non-stop from the Naval Air Station at Patuxent River, Maryland, to Natal, Brazil, with a 13,000-lb payload. On the return flight, the aircraft lifted a record payload of 35,000 lbs over a stage length of 1,200 miles! The following January, a Mars lifted a 20,500-lb payload from California to Hawaii. A modified version, the JRM–1 with a single fin and specially adapted for cargo only was ordered in January, 1945, but as the war was ending, the order was reduced from twenty aircraft to just four, all of which used uprated 2,300-hp Double Cyclone engines. However, the fourth JRM–1, named *Marshall Mars*, set a record on 19th May, 1946, lifting seven crew and 301 passengers from Alameda to San Diego. Even so, plans for a post-war commercial version and for a medical evacuation version (MEDEVAC) were cancelled.

Of course, the Mars and Mariner flying boats were not the only Martin aircraft operating during the war years. Pan American's Martin M–130s were initially on loan to the United States Navy. However, the wartime service of the M–130s, the 'China Clippers', was decidedly unlucky. One of these aircraft, *Philippine Clipper*, arrived over San Francisco in heavy rain while in USN service, some four hours early due to strong tailwinds on a flight from Hawaii on 21st January, 1943. It had flown a holding pattern waiting for the visibility to improve, and while doing so drifted inland. No one is quite sure what happened next, but it is believed that the pilot attempted to drop down through heavy cloud to establish his bearings, and in doing so rammed a hill. The aircraft burst into flames and all aboard were lost. Later that year, in October, *China Clipper* was handed back to Pan American after being replaced by landplanes in Naval Air Transport Service. The aircraft was then put into service on a wartime Pan American route flying to the Caribbean and to the then Belgian Congo, terminating at Kinshasa, then known as Leopoldville, and providing a link in a long and rather circuitous route from the United States to the Middle East a route which at that time had to avoid German and Italian occupied North Africa. On 8th January, 1945, the *China Clipper* took off from Miami for Leopoldville. The first stop *en route* was Port of Spain in Trinidad where, on landing in darkness, the aircraft hit the water heavily and broke up into several pieces. There was no fire, but many of those not killed by the severe impact were fatally injured through swallowing fuel once in the water, with the loss of nine out of the twelve crew members and fourteen out of eighteen passengers.

Martin produced an improved M–130, the M–156, hoping for a Pan American order, and although this aircraft flew successfully, no order was forthcoming from the airline, which had already taken a policy decision to standardize on landplanes for its future development. The solitary M–156, named *Soviet Clipper*, was sold to the USSR, and no more is known about this aircraft's subsequent history, although it may have had some influence on post-war Soviet flying boat and amphibian development.

Before the United States entered World War II, Pan American had received another six Boeing 314s, giving the airline twelve in all, with the first six being named *Honolulu*, *California*, *Yankee*, *Atlantic*, *Dixie* and *American Clipper*. Of the second batch, known as 314As, three were retained by Pan American as *Pacific Clipper*, *Anzac Clipper* and

Capetown Clipper, and the other three were sold to BOAC to become that airline's *Berwick*, *Bangor* and *Bristol*. These aircraft were delivered between April and August, 1941, and after their delivery the original six were also upgraded to Boeing 314A standard, with the addition of accommodation for an extra three passengers, although the same 1,600-hp Wright Cyclone engines were retained. Before the outbreak of war, Pan American had been operating to Marseilles and to Lisbon with the Boeing flying boats, as well as on a series of experimental flights between the United States and the United Kingdom. Once war started in Europe, Pan American operated its first six aircraft in the Pacific, and the second batch of three on the North Atlantic. Meanwhile, BOAC commenced a wartime service with the 314As between Poole and Baltimore, and from Poole to Lisbon, which, as a neutral capital, was an important diplomatic centre. This connected with the Pan American service from the United States.

Almost immediately after the outbreak of war, BOAC had been put onto a wartime footing. The transatlantic services with the 314As, which commenced in May, 1941, were just a part of this. There was also a service through Lisbon to West Africa established early during the war, and landplanes maintained a service between West Africa and the Sudan and East Africa, connecting with the joint BOAC and Qantas flying boat route from South Africa to Australia – a distance which was too great for a direct service and was therefore a giant inverted 'horse shoe' through India and Malaya. After the Japanese entered the war and invaded Malaysia and Singapore, the Empire flying boats used by BOAC and Qantas had to fly 3,512 miles non-stop from Ceylon to Perth, Western Australia.

The flying boats represented a considerable leap forward in comfort compared to many of the other aircraft in use at the time, with the 314As on the North Atlantic service replacing an irregular service by unarmed Liberator bombers in civilian markings, which were obviously vastly inferior in terms of comfort to the 314As, costing BOAC just £259,000 apiece. Flying boats were by this time also a small part of the BOAC fleet, which included a number of bomber aircraft, although always unarmed and with civilian markings, and such dashing aircraft types as Mosquito light bombers, which were used on a special service from Scotland to Sweden, flying over occupied Norway, carrying diplomatic personnel and, to the UK, Swedish ball-bearings. Altogether, BOAC had a wartime fleet of 160 aircraft, which operated 57 million miles, carried 50 million lbs of cargo and 280,000 passengers, over a route network of 54,000 miles.

The sudden outbreak of war in the Pacific naturally caught Pan

American's fleet scattered across the region. *Anzac Clipper*, one of the 314As, was caught in the air between Hawaii and Singapore when the captain received news of the Japanese attack. He immediately turned back, but instead of returning to Pearl Harbor landed his passengers at Hilo, the safest and most south-easterly of the Hawaiian islands, before flying the empty aircraft direct to San Francisco for orders. Another aircraft, *Pacific Clipper*, was at Auckland in New Zealand, and the captain was ordered by the United States Government to take the longest, but safest, route home, returning by way of Australia, India, Arabia, Africa and South America to New York, arriving there on 6th January, 1942, after flying some 35,000 miles. *Pacific Clipper* was requisitioned by the United States Navy and *Anzac Clipper* passed into United States Army Air Force service, being designated C–98s by their new owners. *Yankee*, *Dixie* and *Atlantic Clipper* were also purchased by the United States Navy, but none of these aircraft were actually used by the Navy, while *American*, *Capetown*, *California* and *Dixie Clipper* followed *Anzac Clipper* into USAAF service. The crew of the aircraft remained Pan American employees even while their machines belonged to the USN and USAAF, relieving those services of the need to train C–98 aircrew or absorb them into their seniority patterns. *Honolulu Clipper* remained with Pan American, flying between Hawaii and San Francisco.

The only wartime casualty amongst the Pan American Boeing 314s was *Yankee Clipper*. During the early evening of 22nd February, 1943, on approach to Lisbon after a transatlantic flight, the aircraft's port wingtip touched the surface of the River Tagus as she banked, ready to land, and the whole aircraft swung into the river, killing twenty-four out of the thirty-nine persons aboard.

There were other excitements for these capable aircraft. The USAAF C–98, *Anzac Clipper*, flew round the world on one occasion. The aircraft was initially scheduled to fly to Casablanca to pick up President Roosevelt, Sir Winston Churchill and the Russian dictator Joseph Stalin, and take them to meet the Chinese Nationalist leader, Chiang Kai-Shek, in Australia for a conference. While *en route* from New York to Casablanca, the conference was cancelled and the aircraft was ordered to take a group of US Army officers from Casablanca to Australia, before returning across the Pacific and the United States to New York, flying some 37,000 miles. However, VIPs did use the 314s, and when President Roosevelt flew to Casablanca in January, 1943, he celebrated his 61st birthday while on board the aircraft.

An earlier VIP to sample the 314s was that supporter of naval aviation before the outbreak of World War I, Sir Winston Churchill, who was, of course, also Britain's wartime Prime Minister. In 1942, having visited Washington for a meeting with President Roosevelt, Churchill boarded BOAC's 314A, *Berwick*, at Norfolk, Virginia, for the flight to Bermuda. Churchill was sufficiently impressed to write: 'I travelled in an enormous Boeing flying boat, which made a most favourable impression upon me. I took the controls for a bit, to feel this ponderous machine of 30 tons or more in the air. I got more and more attached to the flying boat.'

So impressed was Churchill that he asked BOAC's captain, Kelly Rogers, whether the aircraft could fly the Atlantic. On receiving an answer in the affirmative, he suggested that the aircraft should fly from Bermuda to England. However, his advisers, Chief of the Air Staff, Air Chief Marshal Sir Charles Portal, and the First Sea Lord, Admiral of the Fleet Sir Dudley Pound, objected, suggesting that Churchill should stick to the original plan to take the battleship HMS *Duke of York* from Bermuda to England on the grounds that this would be safer. Churchill reminded them of the U-boat threat! He won the discussion, as usual. In fact, the worst fears of Churchill's advisers were nearly fulfilled as the British Isles drew near. Lost in fog, the aircraft nearly overflew Brest in France, and then approached the British coast from the wrong direction, so that a squadron of Hawker Hurricane fighters were scrambled to intercept the 'intruder'. Fortunately, they missed their quarry in the fog. In fact, it is unlikely that they would have mistaken the aircraft for a German machine.

Development of shipboard aircraft was nearing an end, for even as the war progressed, the helicopter had made its appearance in Germany and in the United States, while the spread of the aircraft-carrier had meant that no task force or fleet deployment of any size was regarded as satisfactory without a carrier and its air group, providing air cover and reconnaissance which simply could not be matched by the shipboard seaplane or amphibian. One of the few, and relatively limited, improvements was the introduction by the Royal Navy of a successor to the Supermarine Walrus, the Sea Otter, a biplane amphibian in the Supermarine tradition, although with a tractor propeller on its single inter-wing mounted engine rather than a pusher propeller. This aircraft also saw service with the Australian and Dutch naval air arms post-war, but for the most part it was operated as a shore-based search and rescue aircraft.

In the United States, BuAer had decided upon a 'high-speed scout' to replace the aircraft of the SOC Seagull series, introduced during the second half of the 1930s. As might have been expected, the competition was directly between the Curtiss XSO3C–1 and the Vought

The standard Royal Navy shipboard aircraft was the Supermarine Walrus amphibian; shore-based counterparts were used for search and rescue.

XSO2U–1, both of which had to use a 500-hp Ranger engine as specified by BuAer, both of which were single-float monoplanes, and both of which made their first flight during December, 1939. Although the Vought aircraft was considered the superior machine in the competition that followed, the Vought factory was heavily committed to Corsair production as wartime threatened, and so the order went to the Curtiss machine, with an order for 300 for the USN, and another 250 for the Royal Navy under Lend-Lease. The aircraft started to enter service on both sides of the Atlantic in April, 1942, as the SO3C–1, after a number of necessary modifications to overcome problems of stability. The first operational deployment was aboard the cruiser USS *Cleveland*, in the July. Known as the 'Seamew', the Royal Navy rather confusingly called the aircraft 'Seagulls'! Before long, it became apparent that the aircraft was too heavy for the limited power available from its engine, and many of the USN aircraft were converted to wheeled aircraft for carrier operation, cutting the drag of the large float, or to radio-controlled target drones. The Royal Navy's SO3C–2C Seamews were only used for training and drone work, never joining the fleet. Eventually some 800 aircraft were built of all types, many of them landplanes.

A rather more successful aircraft followed, the monoplane Curtiss SC–1 Sea Hawk, the last USN floatplane, and one which enjoyed adequate power, with a 1,350-hp Wright engine, giving it a speed of up to 330 mph. Probably the most potent shipboard seaplane ever to enter service, the Sea Hawk was superior to anything else in service, and although such aircraft were not intended to enter combat, it would have been far superior to the Arado Ar. 196, had it ever met one! Available with wheels or floats, the prototype XSC–1 first flew on 16th February, 1944, with deliveries of the first of 700 aircraft ordered starting in July. The first operational deployment was aboard the heavy cruiser, USS *Guam*. However, when war ended in 1945, the remaining order was cut back, with a total of 568 aircraft delivered and another 384 cancelled by VJ Day.

One other seaplane design, by the float-builder Edo, used a 550-hp Ranger engine and first flew on 28th December, 1945, as the XOSE–1, but war had ended and so had the day of the shipboard seaplane. Worse, the XOSE–1 and a development, the XOSE–2, suffered from aerodynamic problems, so the project was abandoned.

The seaplane in its shipboard form found the end of its usefulness more sudden and more final than any other hydro-aeroplane, with a rapid run down of these aircraft after the end of World War II. The carrier reigned supreme, so cruisers and battleships, the refuge of such aircraft, found their days numbered, and those that survived were soon to become radar- and helicopter-equipped, especially in those navies, such as the Italian and Russian, which were without aircraft-carriers. But, even for the flying boat and the amphibian, the best was past.

8
INDIAN SUMMER

World War II ended with the collapse of Germany and the surrender of Japan in 1945, but the Soviet Union was soon giving her former wartime allies cause for concern as, one after another, the nations of Eastern Europe fell under Soviet domination. Mutual suspicion between East and West soon became the dominant theme in the post-war world, with a cold war developing between the two main power blocs, in the midst of which there were hot spots, mainly the result of Soviet-backed revolutionary movements attempting to unsettle existing governments and alliances. The main areas of friction soon moved away from Europe as the effect of a policy of nuclear deterrence made war in Europe between the North Atlantic Treaty Organization and the Warsaw Pact unthinkable. For the most part, wars were fought indirectly in Africa, Asia and Latin America, and also in the Middle East, where friction between the Arab nations and Israel frequently resulted in short, but bloody, wars between well-equipped adversaries.

For the flying boat, the seaplane and the amphibian, the post-war world was one of change as well, technical change which was to prove to be every bit as drastic as the political changes also taking place. The post-war landplane possessed a performance far in excess of anything in service between the wars, and aircraft for the long over-water intercontinental routes soon became commonplace, some of them having the benefit of turbo-prop propulsion. Others, eventually most, had turbo-jets or, later, turbo-fans, capable of speeds of more than 500 mph. Standards of comfort fell, of course, and soon many were to miss the high standard of service and comfort which had been so highly prized by Imperial Airways and Pan American Airways, even though Pan American World Airways, as it became known before later adopting the now familiar Pan Am, and BOAC both attempted a transfer of such standards to the landplane with the Boeing Stratocruiser. However, this attempt was short-lived and costly. Most of all, the post-war passenger expected speed, and then economy, so that it was no longer possible to establish high standards, though this was becoming less necessary, as flights took less time and included fewer stops. Overnight stops were also dropped as navigational aids and aircraft reliability improved. In 1939, it was regarded as marvellous that the flying boat from Southampton to Sydney should take just nine days, but by May, 1945, an Avro Lancastrian of BOAC would take just sixty-three hours, flying from Hurn, near Bournemouth, and not very far from Southampton, on what remained the world's longest regular air service, at 12,000 miles. Of course, the drop in standards of comfort was particularly acute with the Lancastrian, a converted Lancaster bomber, but better was to come, and journey times were to fall still further.

In the navies and air forces the story was much the same. The growing capability of the long-range bomber had meant that these aircraft, including Boeing Flying Fortresses and Avro Lancasters, were able to take on the rôle of maritime reconnaissance, and later, as the true value of maritime reconnaissance became more fully appreciated, and the capability of the aircraft improved with more modern equipment, specialized aircraft were sought for this rôle. The improvement in the performance of carrier-borne aircraft and the introduction of the aircraft-carriers to navies such as those of Australia, Argentina, Brazil, the Netherlands, India and Canada, meant that the seaplane was doomed. War had not only demonstrated once again the importance of air power, it had forced the building of airfields throughout much of the world, notably in the Pacific and in Africa and Asia, so that it was no longer necessary for airlines or air forces to specify an aircraft which carried its own runway on its bottom. Helicopters took over many flying boat rôles, and especially those of search and rescue and anti-submarine duties. And, to add insult to injury, the earliest experiments with helicopters included experiments with float-equipped Sikorsky R-4s, which could land on a dry surface almost as easily as it could on water.

The return of peace saw several Empires in service; this was TEAL's second aircraft, delivered 1940, sold 1948.

The end came quickly, but not suddenly. For those who wished to do so, there was still a chance to fly in a marine aircraft on some of the major air routes, and on many of the minor ones. While it was clear that the landplane was dominant, a number of factors, including deliveries of new aircraft, training of personnel, and completion of networks of landing fields, meant that there was for a while a reprieve for some of the flying boats. During the war, a number of Short Sunderland IIIs had been modified to carry passengers, mail and cargo for BOAC, and post-war an uprated and much modified development of the Sunderland provided BOAC's Hythe-class flying boats which entered service early in 1946, with later developments. These became known as the Himalaya, Plymouth and Bermuda classes, with the civil Solent flying boat developed from the RAF's Short Seaford, much in the manner that the Sandringham had been developed from the Sunderland III. The Sandringham used four 1,200-hp Pratt and Whitney Twin Wasp radial engines, which gave it a maximum speed of 238 mph, while the Solent used four 1,690-hp Bristol Hercules engines and entered BOAC service in 1948. Passenger accommodation varied, from sixteen or twenty-four on a Sandringham to thirty on a Solent, although after disposal by BOAC, some of these aircraft carried as many as forty-four passengers for their new owners. Even flying boat standards had to change!

Although Qantas resumed limited flying boat operations post-war, the predecessor of Air New Zealand, Tasman Empire Airways (TEAL) was another operator of the flying boat, with Short Solents

One of the civil Sunderland developments post-war was the Solent, seen here on TEAL service.

operating with this airline until their replacement by Lockheed Electra landplanes in 1960. The Royal New Zealand Air Force was another adherent of the flying boat, keeping its Sunderlands until 1960. Meanwhile, BOAC withdrew its flying boats from the London–Johannesburg service on 10th November, 1950, replacing the 4½ day flying boat service with a much faster service by landplanes, and indeed using this route as one of the first for the ill-fated de Havilland Comet I jet airliners. As late as 1966, Ansett, the Australian independent airline, operated two Sandringhams, while in Britain, Aquila Airways operated flying boats until 1958.

Some new airlines were created by the flying boat, including Icelandair of Iceland, which flew some of the few Catalina flying boats to enter commercial service. Yet, in spite of the importance of the flying boat to Pan American for its South American services between the war, one new British airline, the short-lived British South American Airways, did not use flying boats.

Pan American had not, as already mentioned, recovered its flying boats from the USN and USAAF after the end of the war, and nor did the airline look for more modern flying boats or amphibians as replacements. The Boeing 314s suffered the indignity of being passed from airline to airline to work their remaining days. *Capetown Clipper* found herself renamed the *Bermuda Sky Queen* by her new owners, the short-lived American International Airways, and on charter to Air Liaison of London to carry a British delegation to the United States for a United Nations meeting. Unfortunately, flying to the UK the aircraft encountered strong headwinds, resulting in a fuel shortage, so that she was forced down into the sea in bad weather. Still afloat twenty-four hours later, the aircraft was taken into tow by the USCGV *Bibb*, but during the tow the aircraft, riding high on the water and difficult to control as all flying boats were while being towed in the open sea, struck the *Bibb*, and was eventually sunk by fire from the Coast Guard vessel. Fortunately, her crew and sixty-four passengers were rescued before the tow started.

An airline named World Airways bought *American*, *Anzac*, *California*, and *Dixie Clipper* as well as the three ex-BOAC aircraft, again for charter operations. However, the cost of operating these aircraft led to financial difficulties for the airline, in 1949, which resulted in a take-over by new management, and its rapid conversion to landplanes. The airline survives and prospers today! Eventually, all of the Boeing 314s were scrapped.

The Boeings were outlived by another flying boat design first flown in 1942, the three Vought–Sikorsky VS–44s. Although these were intended for Pan American, they entered service with American Export Airlines, which later became American Overseas Airlines. Two of the aircraft crashed, one during a night landing in Latin America and the other during take-off in Newfoundland, leaving one aircraft to be sold to a new airline, Avalon Air Transport, a company which later changed its name to Catalina Air Lines, operating as it did between Long Beach and Catalina Island, a flight of just twelve minutes. The original accommodation for forty passengers was increased to fifty-three, in standards of comfort more appropriate for such a short flight. This aircraft, named *Excalibur*, remained in service on this route until it was withdrawn in 1967 and passed to new owners, ending its days in Peru.

There was also a run-down in the naval use of the flying boat. In the United States Navy between the end of World War II and the outbreak of war in Korea, only six out of twenty maritime reconnaissance squadrons operated flying boats, Martin PBM–5 Mariners, and only nine out of twenty search and rescue squadrons operated amphibians or flying boats, standardizing on Catalinas. In the Royal Air Force as well, the flying boat was being phased out, with Lockheed P–2 Neptune landplanes being obtained on loan from the United States Navy before the RAF took delivery of its own Avro Shackleton maritime reconnaissance aircraft, which had been developed from the Lancaster bomber. In France, the Aéronavale used Lancasters to supplement its Sunderlands.

However, this drift into obscurity did not happen without some form of rearguard action, for the flying boat still had its supporters and

The only one of three Vought–Sikorsky VS–44As to survive was the second aircraft, operated by Avalon Air Transport to Catalina Island, off the California coast.

they were especially successful in the United States at ensuring that modern technology was applied to the flying boat.

A troubled post-war world also offered certain opportunities for the flying boat to show that it still had something to offer. In April, 1949, an RAF Short Sunderland flying boat landed an RAF medical officer under heavy Chinese fire close to the frigate HMS *Amethyst*, trapped in the Yangtze River by the advancing Chinese Communists. When the Soviet Union attempted to blockade West Berlin, RAF flying boats again participated, joining landplanes in the attempt to break the siege by air freighting all of the beleaguered city's needs, flying from Hamburg to the Havel See on the outskirts of Berlin. When the Dutch withdrew from what was to become Indonesia, they bequeathed the new Indonesian Air Force, the AURI, a number of Catalina amphibians.

The Korean War itself was a period of renewed interest in the flying boat. After the invasion of South Korea by North Korean forces on 25th June, 1950, the United Nations forces deployed in the area against the North Koreans found airfields in limited supply, putting much of the emphasis on airpower during the early days of the war on the carrier-borne aircraft of the Royal Navy and the United States Navy, and on flying boats. The Royal Air Force deployed two squadrons of Short Sunderlands to Korea to assist in a blockade of North Korean ports, while the United States Navy used flying boat tenders to support Martin PBM–5 Mariner and P5M Marlin flying boats, which were also operated on blockade duties. Another effect of the Korean War was the rearmament of Japan, although the new Japanese armed services were rather diplomatically termed 'self-defence forces'. However, when the Japanese Maritime Self-Defence Force was reformed in July, 1954, amongst its earliest aircraft were numbers of Convair PBY–6A Catalina and Grumman SA–16 Albatross amphibians, the latter being used for search and rescue while the former undertook maritime reconnaissance duties.

A new look flying boat came with the Martin XP5M–1 Marlin, which was flown for the first time on 30th May, 1948. Based on a new hull form, the prototype used the wings of the PBM–5 Mariners with twin 2,700-hp Wright Cyclone engines in an aircraft which showed a marked change in appearance from earlier flying boats, with a longer, narrower hull, with a single step introduced to the keel line and a planing bottom, a high tailplane and hydroflaps for quick turns while taxiing. In common with World War II aircraft, the XP5M–1 included gun positions in the nose, tail and mud-upper positions. The first of four pre-production aircraft, designated the P5M–1 Marlin, flew on

E2

Martin PBM–5 Mariner flying boat being lifted aboard the seaplane tender USS *Salisbury Sound* in Japanese waters during the Korean War.

Martin P5M Marlin flying boat refuelling from a ship.

The clean modern lines of the Marlin in flight.

22nd June, 1951, using 3,200-hp Cyclones and fixed floats rather than the retractable wingtip floats of the prototype. In this version the aircraft was equipped with APS–44A radar, and torpedoes or depth-charges could be carried in the engine nacelles.

During the Korean War, 110 P5M–1s were built for the United States Navy, and a further seven P5M–1Gs were built for the United States Coast Guard, bringing a useful 270 mph maximum speed and a 1,360-mile radius of action to both maritime reconnaissance and search and rescue operations. An improved version, the P5M–2, was flown for the first time on 29th April, 1954, introducing a 130 million candle-power searchlight in the starboard wing and a MAD 'stinger', a magnetic anomaly detector to detect changes to the earth's magnetic field made by a submarine, in the tail. The USN received 117 P5M–2s between 1954 and 1960, while the USCG received just four P5M–2Gs, but by the late 1960s, there were ten Marlin squadrons and twenty Neptune landplane squadrons in the United States Navy, leaving the landplanes clearly in the lead. In an attempt both to fill a gap in Australia's maritime defences and stimulate renewed interest in the flying boat, ten Marlins were loaned to the Royal Australian Air Force in 1959, but only for a short period. The P5M–1 and P5M–2 were redesignated SP–5A and SP–5B in 1962, when all of the US armed forces introduced a common set of aircraft designations for the first time. During its service life the Marlin was provided with more up-to-date submarine detection equipment, including Jezebel sonobuoy systems, which the aircraft would drop into the sea in an attempt to listen for submarines. Any signals would be transmitted back to the aircraft.

However, all good things come to an end, and it was an SP–5B of VP–40 which made the last flight by an operational USN flying boat on 6th November, 1967, taking off from San Diego, close to the spot of Curtiss's first seaplane flight on 6th January, 1911.

During the operational life of the Marlin, attempts were still being made to re-establish the flying boat as a significant aircraft type for the United States Navy, with further designs from both Convair and Martin. Convair developed a turbo-prop-powered flying boat, the XP5Y–1, using four 5,100 shp Allison T-40 engines, a more modern fuselage and hull form, and thin low-drag wings, although this last feature created other problems, since it was impossible to use retractable wingtip floats with wings with such a thin cross-section, and this left the floats to create drag in flight! The first flight of the one and only XP5Y–1 was on 18th April, 1950, at San Diego. Amongst the modern features on the prototype aircraft were five remote-controlled gun turrets, each with twin 20-mm guns, and the aircraft could carry up to 8,000 lbs of ammunition, a significant advance over earlier flying boats. The planned second XP5Y–1 was never completed and the first aircraft crashed on 15th July, 1953. A transport development, the R3Y Tradewind, was nevertheless put into limited production, with eleven being built for the USN, although these were retired in 1958 due to technical difficulties.

A design competition in October, 1952, took place between Martin and Convair as both companies sought a contract to build a jet-powered flying boat for the United States Navy. Martin had earlier built its P4M–1 Mercator compound flying boat, using a mixture of piston and jet propulsion, but this had not entered volume production, as indeed few compound aircraft ever did, even though some basically piston-engined designs received either jet or rocket booster engines to augment performance, especially when taking off from high-altitude airfields. The Martin design won the jet flying boat competition, with the Martin Model 275 being built to fulfil the USN's requirements not for a maritime reconnaissance aircraft but for a fast mine-layer. The first of two XP6M–1 Seamaster prototypes flew on 14th July, 1955, and was followed by the second aircraft on 18th May, 1956. The prototypes used 13,000-lb thrust Allison turbo-jets with afterburning, although production aircraft used four 17,500-lb Pratt and Whitney J75 turbo-jets, which gave a maximum speed of more than 600 mph and a 1,500-mile operational radius. A sleek aircraft with a narrow fuselage and a 'T' tail, the Seamaster dispensed with wingtip floats by having pronounced anhedral on the wings, so that wingtips themselves provided the stabilizing action by touching the water. The bomb bay

A rare turbo-jet flying boat, the Martin YP6M–1 Seamaster in flight.

used a watertight rotary door.

Long patrols on the Seamaster must have meant a marked improvement in comfort for the crew, with this being one of the few pressurized flying boats built. Although both the prototypes were lost in accidents, a pre-production YP6M–1 was built and flown on 20th January, 1958, with a further five pre-production machines following shortly afterwards. Whether the production order was reduced due to second thoughts on the rôle of the flying boat by the USN, or whether only a few aircraft were regarded as necessary since the operational plan for the aircraft anticipated the Seamaster being deployed in small groups, widely scattered and refuelling from submarines, the production order called for just twenty-four aircraft. Worse was to follow, for after the first flight of a production P6M–1 on 3rd March, 1959, the production order was reduced to just four aircraft, and by August, 1959, these were withdrawn from service.

While the performance of the Seamaster was undoubtedly sparkling for a large aircraft, the power offered by the turbo-jet engine was of obvious potential for fighter aircraft. However, only two jet-powered hydro-aeroplanes were ever designed as fighters, one British and one American, and neither entered production.

The more interesting design was undoubtedly the American project, the Convair XF2Y–1 Sea Dart, intended to operate from coastal waters, bays and lakes. A delta-wing aircraft, not unlike Convair's landplane fighters of the day, the Sea Dart floated on a watertight belly – rather than floats or a flying boat hull – until take-off speed lifted the aircraft onto retractable twin skis. Two prototype aircraft were ordered by the United States Navy in 1951, but after inspecting a full-

Take-off for the Convair XF2Y–1 Sea Dart jet fighter.

sized mockup the following year, a production order for twelve was issued, and this was later increased to twenty-two aircraft, all before the aircraft made its first flight on 9th April, 1953. The production contract was then cut from twenty-two aircraft to just five in March, 1954, including the prototypes, the second of which was destroyed on 4th November, 1954, while a pre-production YF2Y–1 made its first flight on 4th March, 1955, but was lost. The last two aircraft never flew.

Originally fitted with a Westinghouse J34 engine, the aircraft were later fitted with a 5,725-lb thrust J46, which provided a speed of Mach 0.99 in level flight, while the sound barrier could be broken in a shallow dive, making this aircraft the world's only supersonic seaplane. The cancellation of the programme was largely due to the removal of the requirement for a waterborne fighter, but the hazards of high-speed take-off, with the damage likely from flotsam and jetsam increased by the speed of the machine, must also have been a deciding factor.

A more conventional, if that can ever be the right word for a jet-powered hydro-aeroplane fighter, approach was the Saunders–Roe SR.A1 jet-powered flying boat, which in fact preceded the Sea Dart. The project originated with a British Air Ministry requirement for a flying-boat fighter able to operate from coastal waters in the Pacific, but the war ended before the aircraft flew, thereby denying the chance to test this novel concept operationally. Designed by Henry Knowler, the aircraft used twin 3,250-lb thrust Metrovick Beryl turbo-jets mounted side by side within the fuselage, with a nose air intake. It was amongst the first aircraft to provide the pilot with an ejector-seat. There were three prototypes, with the first flying on 15th July, 1947. The second, with uprated 3,500-lb thrust engines, flew in early 1948. Still more powerful engines, of 3,850-lbs thrust, were fitted to the third aircraft. The second aircraft was lost in an accident, but basically the aircraft performed successfully, and once again it was the withdrawal of the requirement which finally saw an end to the project. It was already clear that jet fighters could be operated successfully from aircraft-carriers, helped by steam catapults, the angled flight deck, the mirror deck-landing system, and by larger ships, with the first three items being British inventions. Good carrier-borne aircraft were the real downfall of the seaplane fighter, or the flying-boat fighter for that matter! Unlike the Sea Dart, which had a fairly respectable turn of speed, the SR.A1 was essentially a 400-mph aircraft, and even some piston-engined fighters could do as well!

The quest for size had not been forgotten, however, and again there were British and American designs for large long-range flying boats, though these were far too large for the air transport market of the day.

Saunders–Roe, or Saro as it was more usually called, was in fact the source of the largest British flying boat, the Princess, on which design work started in 1943, although the war prevented construction work from being started. The idea was that the aircraft would be a transatlantic flying boat for post-war use by BOAC to operate between Southampton and New York, carrying 105 passengers. This figure was rather on the low side for the size of the aircraft, so obviously pre-war standards of comfort and service were very much in mind. A prototype was ordered by the Ministry of Supply in May, 1946, originally planned to be powered by twelve Rolls-Royce Tweed turbo-props in six nacelles, with the engines mounted side by side and each pair driving a contra-rotating propeller. Later on though, the aircraft was built using ten Bristol Proteus turbo-props, still in six nacelles, but with the outer nacelles having only one engine each. The gross weight of the aircraft was 315,000 lbs, 20,000 lbs more than the giant Bristol Brabazon landplane, and was powered by 3,780-shp Proteus engines. With a span of 220 feet and a length of 148 feet, it was still expected to cruise at 360 mph and have a range of 5,270 miles. Even before the first flight of a Princess, the customer, BOAC, decided to change to landplanes, but work continued, with the first prototype flying on 22nd August, 1952, while the other two prototypes were cocooned at Calshot, until scrapped in 1967. The aircraft performed well on trials, but eventually was laid up and finally scrapped. At intervals, plans had been announced for the formation of new airlines anxious to use the aircraft and bring back the flying boat era, but nothing ever came of these plans. Certainly, Saro were sufficiently discouraged to scrap plans for the Duchess, a jet-powered flying boat with seventy-four seats and powered by six 5,000-lb thrust de Havilland Goblin turbo-jets.

Nevertheless, the distinction of building the largest flying boat must go to the American aviator and millionaire eccentric Howard Hughes. The aircraft was intended to be a massive military transport flying boat. It had a 320-foot wingspan and a 218-foot-long fuselage, compared with the 195-foot wingspan of a Boeing 747 jumbo jet! Hughes started work on his flying boat in 1942, but the wartime shortage of materials encouraged him to use laminated plywood, resulting in the nickname of 'Spruce Goose' for his aircraft, which nevertheless also enjoyed some $40 million of US Government funds. The government lost interest once World War II ended, but Hughes was determined to complete the project, which he had designated the

Hughes Hercules. In part, his determination was provoked by criticism of the project by American politicians, who claimed that the aircraft would never fly. The eight-engined flying boat was launched at Long Beach on 2nd November, 1947, and, with Howard Hughes at the controls, the aircraft immediately started its taxiing trials, showing good handling on the water, before starting a take-off run and flying for almost a mile at some twenty feet above the water and at less than 100 mph. This was the one and only flight of the Hercules, after which it was put into storage, leaving its critics to debate whether or not it could fly properly, even though the old minimum criteria for flight laid down in the days of the pioneers had been exceeded. It was certainly a worthwhile achievement for a first outing, but not one which would have earned the aircraft its keep in any late-1940s airline or air force!

However, not all of the post-World War II projects were doomed to failure, and not all were vain attempts at speed or size, for there were aircraft, mainly amphibians, which enjoyed commercial success, and which were to extend the usefulness of the hydro-aeroplane for many years, until its limited rebirth in Canada and Japan during the early 1970s.

One of the most successful post-World War II amphibian designs, and certainly the most successful in the West, was the Grumman HU–16 Albatross, developed from the Grumman Mallard, an early post-war project, designed for commercial operators which became the first post-war Grumman aircraft. The Albatross first flew in 1947, primarily as a search and rescue amphibian using Wright R–1820

A rare success for the amphibian post-war was the Grumman HU–16A Albatross.

A USAF search and rescue Albatross in flight; the 'step' can be seen clearly below the retracted wheel.

engines. The aircraft saw service with the USCG as the HU–16E, the USN as the HU–16C and HU–16D, and as the HU–16A and HU–16B in the USAF service. The original designation of SA–16 for the USAF and UF–1 for the USN having been changed in 1962 when all US service aircraft designations were standardized for the first time. The early SA–16A, in USAF service, was soon followed by the uprated SA–16B, and subsequently most aircraft were converted to this standard, with a higher weight and improved performance from more powerful 1,425-hp versions of the Wright R–1820 engine, giving the aircraft a maximum speed of 240 mph and a range of up to 3,000 miles or a payload of up to 10,000 lbs. One of the most successful Grumman designs, the Albatross was exported to Argentina, Brazil, Chile, West Germany, Indonesia, Italy, Japan, the Philippines, Portugal and Taiwan in its basic search and rescue version, while an anti-submarine development entered service in 1961 with the armed forces of Norway, Canada and Spain. Even today, almost thirty years after its first flight, the Albatross remains in service in limited numbers with the Hellenic Air Force, and in the Philippines, Thailand, Taiwan, Peru and Venezuela. The Albatross can operate from snow or ice using special runners.

Two other less successful amphibians included the small Piaggio P.136, and the Short Sealand. The P.136, a high-wing monoplane with twin 260-hp Lycoming engines driving pusher propellers, was used by the Italian armed forces for coastal patrols, including search and rescue, and for communications duties, having a range of just 400 miles and a maximum speed of 180 mph – although this performance is

Too late for the flying boat,
the Short Shetland never entered RAF service.

Relatively limited production was achieved by the attractive Short Sealand amphibian.

fairly good for the power available. Shorts, who had been less than successful with their Shetland flying boat for the RAF, also developed an eight-seat amphibian, the Sealand, which the company produced in small numbers. The Indian Navy was one of the few customers.

As purpose-built seaplanes and flying boats became more of a rarity, attention turned again to producing seaplanes or floatplane conversions of landplanes. There were also attempts to reduce take-off distances, including the 'pantobase' experiments with USAF transport aircraft during the mid-1950s, but the growing size and reliability of the

Another small volume production post-war amphibian was the Piaggio P.136.

The first of a line of successful utility aircraft from de Havilland Canada was the Beaver, which could be fitted with floats or wheels.

helicopter and of short take-off aircraft meant that such experiments provided little practical development.

However, one of the most successful manufacturers of short take-off transport aircraft, de Havilland Canada, was also the most persistent manufacturer of aircraft in seaplane or amphibian form from their normal landplane configuration. Established as an 'overflow' factory by the British de Havilland concern during World War II, post-war DHC began to produce a range of aircraft of its own, and the company was in fact acquired by the Canadian Government during the late 1970s, long after de Havilland had been merged into Hawker Siddeley Aviation, and that in turn into British Aerospace.

DHC's first design was the DHC–1 Chipmunk, a small landplane trainer, but the next design, the DHC–2 Beaver light transport and communications aircraft, which first flew on 16th August, 1947, became the first of a line of aircraft to be offered in seaplane form, and occasionally amphibian form, when customers demanded it. The Mark 1 version of the Beaver used a single nose-mounted 450-hp Pratt and Whitney Wasp Junior radial engine, for a maximum speed of about 180 mph, or 155 mph as a seaplane or amphibian. Altogether 1,657 were built, although only a small number were the hydro-plane conversions with Edo floats. Subsequent development of the Beaver including the fitting of a Pratt and Whitney PT6A turbo-prop to create the Turbo Beaver, and the Otter, a stretched development with seats for up to twenty passengers in the Twin Otter turbo-prop development, compared with the six passengers and two crew of the

The DHC Twin Otter can also be found on occasion with floats, as with this Peruvian Air Force example.

original Beaver. More than 800 Twin Otters have been built, the vast majority as landplanes. These aircraft are available as floatplanes, although DHC's larger aircraft, the Caribou, Buffalo and the new Dash 7 and Dash 8, are available only as landplanes.

Post-war, the Soviet Union maintained its interest in the flying boat for maritime reconnaissance operations, using aircraft from the Beriev design bureau.

The first of Beriev's post-war designs, which were far more successful than his pre-war aircraft, was the Be–6, a twin-engine gull-wing flying boat codenamed 'Madge' by the NATO countries: the practice being to use codenames with the letter 'M' for marine aircraft, a useful practice given the long delay before official Soviet designations become known. The two 2,000-hp Shvetsov ASh–73 radial engines provided a maximum speed of 250 mph and range of up to 3,000 miles, while modern anti-submarine and anti-shipping equipment included a retractable radome – some aircraft were fitted with magnetic anomaly detectors – and bombs, depth-charges, mines and torpedoes could be carried under the wings. The aircraft was phased out during the early 1960s. A development of the Be–6 was the Be–12 amphibian 'Mail', also known to the Russians as the Tchaika, using twin 4,150-shp Ivchanko AI–20 turbo-props, giving a maximum speed of 340 mph and a range of some 3,000 miles, although this aircraft also incorporated a number of hydrodynamic improvements to the hull form. First appearing in 1961 at the Tushino Air Display, the Be–12 was accompanied by a twin-jet flying boat also from the Beriev bureau, the Be–10 'Mallow', with a very modern hull design and two 14,330-lb thrust AL–7PB turbo-jets, giving a maximum speed of up to 566 mph and a range of some 4,000 miles. Both of these aircraft could carry anti-shipping missiles under the wings, in addition to more conventional weapons, and both established a number of speed records, including some with payload to altitude, with the Be–12 lifting 22,000 lbs to 36,680 feet on one occasion, and the Be–10 lifting 33,000 lbs.

However, for many years the flying boat and amphibian were to be virtually neglected, leaving a small market unserved, or served by aircraft which were becoming increasingly elderly, while other aircraft types with shorter take-off, or with vertical take-off in the instance of the helicopters, took over in those parts of the world in which airports were still few and far between. Many of the short take-off aircraft were more economical than the amphibian or flying boat, and all were safer, while the helicopter, even though it was the most expensive aircraft of all, offered many other advantages, especially for search and rescue and for anti-submarine duties.

9
THE FLYING BOAT TODAY

Between the two wars few air forces, naval air arms or army aviation units were without some form of hydro-aeroplane, be it a flying boat, an amphibian or a seaplane. Airlines operating over the longer routes tended to be flying boat or amphibian operators, with the latter aircraft sometimes being selected, as in the case of Pan American, to make emergency landings possible when flying over islands *en route*. Of the world's great airlines, only KLM remained aloof from the flying boat. A number of leading aircraft were available with float-conversions, and although the most successful transport aircraft of all time, the Douglas DC–3 or C–47, was only in one instance completed as a seaplane, nicknamed the *Gooney Bird*, its German rival, the Junkers Ju.52/3M, the second most significant airliner, was not uncommon in its seaplane form, especially with operators such as Finnair. In a post-war world in which maritime reconnaissance had become vastly more important and, thanks to modern equipment, considerably more effective than it had been between the wars or during either, and in which long-distance air travel had increased tremendously, the flying boat was unable to benefit. A boom in the construction of large and longer-range aircraft there was, but it passed the flying boat and its relations by.

The main reason for this, of course, has to be the growth in air travel which has made the cost of building airports and runways worthwhile, and once most of the world does it, anywhere planning to become a tourist centre, for example, has to follow suit since no one is really going to invest in flying boats just to operate to a particular destination! Even the helicopter in its commercial form has suffered from this attitude, with success being self-defeating since it only proves the case for investment in runways. World War II also helped, ensuring that in the immediate post-war world, there were many new airports and landing fields, and some of the less accessible places proved to be the most interesting from the point of view of the tourist trade.

The reason why conventional aircraft are preferred to flying boats or helicopters is simply that they are faster and more economical. Speed reduces journey times, improves competitiveness and increases aircraft utilization. Both the flying boat and the helicopter need far more power to lift a given load over a stated distance than a conventional aeroplane, and both have additional maintenance requirements. For the flying boat, the amphibian and the seaplane, additional safeguards have to be built into maintenance programmes because of the increased risk of corrosion due to the aircraft being subjected to salt water. Also, flying helicopters or marine aircraft requires greater skill on the part of the aircrew.

This bring us to another point, safety. Mention has already been made of the problems of getting the aircraft to unstick and the dangers of porpoising during take-off and landing, while the difficulty in assessing the aircraft's height above the water during the landing approach is another source of danger. There are other factors. Approaches over water generally mean fewer of the navigational aids generally taken for granted by air travellers today are available – this is also a criticism of coastal airport sites – so that safety is reduced. The risk of bird-strikes and fog is increased. And passenger escape from an aircraft which crash lands as a result of under- or over-shooting a runway on dry land is generally far easier than when the aircraft has struck water heavily. The very poor weather in which modern landplanes can operate safely provides greater reliability for the air traveller and a more effective anti-submarine deterrent effect in maritime reconnaissance operations than the hydro-aeroplane can offer. Take-off by a hydro-aeroplane in a heavy swell is difficult and dangerous, even when possible.

The large numbers of passengers, occasionally in excess of 500, and the cargo and baggage accommodated in a modern airliner would also be more difficult to handle with an aircraft floating, constantly moving with the water, than one sitting still on the ground. Ships cope because they are larger and draw more water than the relatively light aircraft, and of course because they are often loaded from above, while ferries offer a single major loading point which can be secured in a way that loading an aircraft other than a military freighter could not be. Then too, without taking on large quantities of ballast, the loading of thirty or forty tons of cargo and baggage above the waterline, followed by three or four hundred passengers, could easily create a dangerous top-heavy aircraft.

These practical limitations may sound defeatist, but it is important to appreciate just why the flying boat, the amphibian and the seaplane are no longer with us. The criticism is often made that landplanes need several thousand feet of runway, indeed two miles and more is not unusual or excessive from the operational point of view, but if the traffic of a major airport such as Heathrow were to be handled by

The Canadair CL–215 amphibian shows that such aircraft still have a vital rôle to play, water-bombing.

hydro-aeroplanes, that would effectively mean closing off several miles of waterway to other users, building massive terminal buildings and aircraft handling facilities on a coastal site, and ensuring that people in an area close to the coast would not suffer unduly from the aircraft noise generated by such an airport. It would be sensible to assume that such an airport would be located in a spot with plenty of water and shelter; the Solent between Hampshire and the Isle of Wight could be one place, at least for the landing and take-off areas, since Southampton Water would be unable to handle modern aircraft movement levels without being closed to shipping. But the noise would affect those on the Isle of Wight from Yarmouth to Ryde or even Bembridge, and those in Hampshire from Southsea or Gosport to Lymington. It would also still be less convenient for most travellers than Heathrow or Gatwick.

Sentiment alone cannot justify the hydro-aeroplane, and an impractical approach can do nothing to save it in a world in which economics and safety must be the deciding factors.

Yet, in spite of this, there are still areas of the world in which such aircraft thrive, albeit in a far more modest way than in the past. In 1983, some ten air arms operated flying boats or amphibians, while a few more operated aircraft such as the de Havilland Canada Twin Otter equipped with floats. Three air transport operators maintain flying boats or amphibians, including one operating a PBY–5 Canso in Canada, an Indonesian operator of a Grumman Albatross and three PBY Catalinas with an airline in Chile. These aircraft are generally used for charter work, and again, a number of small third level or commuter airlines also continue to operate float-equipped Twin Otters. Some of the airlines operate landplanes as well, including the Indonesian operator of the Albatross, Airfast Services. However, one of the third-level operators is also worthy of note, Chalk International Airlines, which has a long history of operations between Miami and the Bahamas. Today, the fleet consists of Grumman Mallards, of which there are four, converted to turbo-prop propulsion and each carrying seventeen passengers, and three Grumman Albatrosses, converted to airliner configuration from the standard HU–16 utility aircraft. The other operators of such aircraft include the Hellenic Air Force, with eight Grumman HU–16B Albatrosses and ten Canadair CL–215 amphibians, the Soviet Navy with more than eighty Beriev Be–12 'Mail' amphibians, while a large number of the old Beriev Be–6 'Madge' flying boats remain in the Communist Chinese Navy. The Indian Navy has five HU–16B Albatrosses, the Philippine Air Force has four, the Taiwanese eight, the Peruvians four and the Venezuelans

A Spanish Canadair CL–215 in flight.

four, while the Royal Thai Navy has two HU–16Bs and two CL–215s. Outside the Communist countries, the largest operator of flying boats today is the Japanese Maritime Self-Defence Force, with nineteen Shin Meiwa PS–1 maritime reconnaissance and seven US–1 search and rescue flying boats. Spain and Jugoslavia have introduced CL–215s fitted with radar for coastal patrols, and the Thai aircraft are equipped in this way.

Other non-military operators of such aircraft include government departments using Canadair CL–215 amphibians for fighting forest fires, in Canada and France in particular. This new fire-fighting rôle for the flying boat started after World War II, with Catalina flying boats undertaking this duty and offering rather faster topping up of the water tanks than a landplane by scooping up water during a 'touch and go' landing and take-off. The future of this type of work became questionable as the aircraft began to age, while their payload was relatively limited – indeed, in some countries, large helicopters were even tried, especially in the Soviet Union. Yet, in Ontario, Canada, one of the areas worst affected by forest fires, it was estimated during the early 1960s that the presence of a fleet of thirteen water-bombers could have saved $25 million of lost timber and property over a three-year period – let alone the risk to life. And of course the figures, high though they may seem, were at the then current prices – inflation, as well as fire, will have taken its toll!

Given this background, it was not surprising that it was an Ontario-based aircraft manufacturer, Canadair, the old Canadian subsidiary of Vickers, which came up with the answer: a completely new, purpose-built water-bomber amphibian. While the aircraft was intended to be available for other rôles, such as utility, cargo, passenger-carrying and search and rescue, the design was first and foremost one for a water-bomber. First flown in September, 1967, the Canadair CL–215 is a high-wing monoplane of relatively simple and traditional appearance, with little streamlining and with two 2,100-hp Pratt and Whitney R2800 radial engines. A wingspan of 93 feet 10 inches and a 65-foot-long fuselage enable the aircraft to lift 12,000 lbs of water or, as an alternative, accommodate up to thirty-two passengers in its commercial form, or nineteen passengers in a cargo and passenger utility configuration. A maximum speed of just 185 knots, about 210 mph, and a maximum range of 1,130 miles complete the picture of an aircraft which was designed for durability and ease of maintenance, and which uses piston engines for greater fuel economy, since the design specification envisaged that most of the aircraft's flying would be at low altitude and frequently on short hops from lakes, discharging water over a fire, and then returning for more water. Provision was made in the design for rapid conversion to the utility rôle, since water bombing tends to be a seasonal business. The aircraft takes on its water load while planing, or running, on the step, and taking some 1,200 gallons of water aboard in a 5,000-foot run, and dropping this at about 100 mph over the fire. The amphibian retains its value in this rôle because of the existence of large stretches of forest in remote areas of Canada, France and Spain, amongst other areas of the world, where fire-fighting teams find access to the area difficult and time-consuming when time is of the essence. The water load of the CL–215 is half as much again as that of the Canso or Catalina, and while simplicity is inherent in the design, attention was given to reducing problems such as porpoising, although collecting water on

A maritime patrol and search and rescue Canadair CL–215 of the Royal Thai Navy.

Take-off for a Shin Meiwa PS–1 turbo-prop flying boat of the JMSDF.

the step means that the aircraft is planing at high speed at just below rotation speed.

The coastal patrol variants of the CL–215 are equipped with a nose-mounted AVQ–21 weather and search radar. Although designs exist for passenger-carrying developments, it is a sign of the development of other aircraft types that this version has not been ordered at the time of writing.

Larger than the CL–215, the other hydro-aeroplane in production today is the Shin Meiwa PS–1, produced by the successor to Kawanishi, who undertook the initial development work, basing the original design on that of the Grumman Albatross series. However, the PS–1, with its four 2,850-shp Ishikawajima-built General Electric T64 turbo-props, is a far larger aircraft, with a 108-foot-8-inch wingspan and a length of 110 feet. A number of refinements have been incorporated in the design to provide a good planing hull and short take-off and landing characteristics, for although an Albatross was modified to serve as a flying scale model for the aircraft, the PS–1 is a flying boat rather than an amphibian. Two prototype aircraft, designated PX–1, were designed and developed after work with the Albatross, and the first of these flew for the first time on 5th October, 1967, while the second flew on 14th June, 1968. The Japanese Maritime Self-Defence Force then received two pre-production aircraft for evaluation during 1972. The PS–1 has a maximum speed of 340 mph, although its range is only some 1,400 miles, but the aircraft has the very low landing speed of just 50 mph. It also has a full complement of anti-submarine equipment, including a magnetic anomaly detector, and can be fitted with torpedoes. The JMSDF operates nineteen PS–1s and seven US–1s, with small numbers of aircraft being ordered from time to time, but the main JMSDF anti-submarine capability is, as one has to expect today, based on landplanes, with the flying boat providing some additional maritime reconnaissance support, especially in areas remote from airfields with runways long enough for larger aircraft. While commercial versions of the PS–1 series have been mooted from time to time, again there have been no orders.

The somewhat restricted market for the flying boat has not prevented one of the most famous flying boat manufacturers from returning to this type of aeroplane. No less a company than Dornier is working on a development of the Dornier Do.24TT, using an aircraft employed until the early 1970s in a search and rescue rôle by the Spanish Air Force. There are differences, however, since the prototype of what may still become a new line of flying boats uses twin Pratt & Whitney Canada PT6A–45 turbo-prop engines and a completely new wing, an enlarged version of the wing used on Dornier's successful Do.228 regional airliner, a landplane. While the aircraft on which Dornier is working is intended to be a flying test-bed for the engines and the wing, known as the TNT or 'tragflugal neuer technologie', Dornier's longer-term ambitions centre around a series of amphibians rather than a flying boat. Indeed, the aircraft fuselage

Dornier's experimental Do.24TT takes to the air. The new wing shape can be seen, with the addition of sponsons and three new Pratt & Whitney PT6A–45B 1,125-shp turboprop engines.

could undergo some redesign before a new generation of Dornier maritime aircraft enters production, making the most of modern materials and the advances in hull design which have been made since the days of the Do.24. In fact, Dornier is one company which believes that there is still a significant market for the flying boat, basing this on the existing number of elderly aircraft still in service, although one might question whether that is a big enough market, and whether the operators could afford to buy brand-new aircraft.

The overwhelming victory of the landplane for maritime reconnaissance and airliner use should not leave one believing that there are no opportunities left for hydro-aeroplanes. Even a few aircraft with water-borne capability can be an added insurance in a defence force, making search and rescue, maritime reconnaissance and transport operations not only that little bit more flexible but providing a fall-back capability in case airfields are attacked and runways rendered unusable, or in the case of the unexpected. A good example lies in the problems encountered by the British Task Force charged with the removal of Argentinian forces from the Falkland Islands. The open reaches of the South Atlantic would not have made good flying boat territory, but the point is that even after the islands were retaken, Royal Air Force maritime reconnaissance aircraft could not be based there due to the shortness of the runway at Port Stanley. Had flying boats or amphibians been available, they could have been deployed as soon as the British forces secured a bridgehead at San Carlos in the northern end of the Falklands Sound, and been operated from that time onwards. They would also have made protection of Thule and the South Sandwich Islands easier, inhibiting Argentinian raids or landings on these remote islands and sparing the need to send frigates or destroyers which have a crew of between 200 and 300 men, and which sail at just 30 knots – this is the advantage of using large aircraft for maritime reconnaissance: the ability to cover a wide area quickly with minimal manpower, while surface vessels provide intensive cover, especially when equipped with heliçopters, over a small area which needs greater security than the maritime reconnaissance aircraft can provide. The presence of even a handful of transport flying boats would also have made the job of supplying forces in the Falklands easier than it proved to be.

It may still be that given the growing sophistication and effectiveness of anti-runway, or 'runway denial', weapons the flying boat will have a rôle to play in the navies and air forces of the world. However, there is strong competition, and increasingly well-established competition at that, from the short take-off airliners and transport aircraft which are currently on the market, even though these aircraft are, without exception, short or, at the most, medium range. The possibility of a come-back, or of the Japanese and Canadian acceptance of these aircraft spreading, is remote, but not completely impossible. Unfortunately, the same cannot be said about the seaplane or the fighter and reconnaissance hydro-aeroplanes, whose usefulness has been completely eclipsed by the brilliance of the British Aerospace Harrier and Sea Harrier vertical take-off fighter aircraft.

Perhaps we should remember today, when the world flying boat and full-sized amphibian population totals only some 200 or so aircraft, that the market for these aircraft was never really big. There were the exceptions, the Catalina and the Sunderland being prominent amongst them, but a production run of just three Martin M-130s, or even twelve Boeing 314 and 314A flying boats, pales into insignificance against some 600 Boeing 747 Jumbo jets, and another 600 odd aircraft of Lockheed TriStar and McDonnell Douglas DC–10 size. Perhaps if the demand for long-distance flight had been what it is today, landplane development between the wars would have kept pace, and attention would have been given earlier to the problem of building better undercarriages and more airports where they were needed. The truth is that not until World War II did these aircraft ever throng the skies – they were for the most part elusive and exclusive, and part of an age when flying was fraught with thrills and danger.

In much the same way as the British Marine Air Navigation airline sought to operate from Southampton to Jersey and Cherbourg, bridging the English Channel, Boeing returned to fast, water transport during the late 1970s with their Jetfoil, in effect a type of hydrofoil, but with the difference that the foils were not water piercing but were below the surface of the water, providing a smoother ride. The Boeing Jetfoil was described in the publicity of some of its operators as the 'boat that flies', suggesting some confusion with the flying boat. Although operated successfully in Washington State and British Columbia, in the Hawaiian Islands and the Far East, the Boeing Jetfoil has had mixed fortunes on the English Channel and the Irish Sea. The Jetfoil suffers from problems with its wave-sensing equipment in confused seas or when the waves are coming from either port or starboard rather than from directly ahead. It is also more expensive to operate than an aeroplane, and the market is limited to finding suitable opportunities for services between important cities on either side of a suitable stretch of water – of which there are not that many within the Jetfoil's range. However, the Jetfoil is more of an example of high-speed water transport than of flying in any proper sense of the word. It is in its infancy, struggling for acceptance in a world in which the existing forms of conventional transport are already highly developed.

There are still small flying boats for the private owner, such as the Thurston Teal, and some light aircraft designs, including the Cessna 150 and 170 series, lend themselves to being float-equipped – ideal for private owners in many parts of the world where a lake or a bay is a more convenient aerodrome than any airfield. But these aircraft obviously lack the certain something which characterized the flying boat or the Schneider Trophy seaplanes.

Bibliography

Allward, Maurice, *An Illustrated History of Seaplanes and Flying Boats* Moorland, 1981

Casey, L. S., *The Illustrated History of Seaplanes and Flying Boats* Hamlyn, 1980

Duval, G., *British Flying Boats: A Pictorial Survey* Barton, Truro, 1973

Duval, G., *World Flying Boats: A Pictorial Survey* Barton, Truro, 1975

Duval, G., *British Float Planes: A Pictorial Survey* Barton, Truro, 1976

Duval, G., *World Float Planes: A Pictorial Survey* Barton, Truro, 1977

Knott, R., *The American Flying Boat: An Illustrated History* Conway Maritime Press, 1979

Munsen, Kenneth, *Flying Boats and Seaplanes since 1910* Blandford, 1971

Rance, A. B., (edited by) *Seaplanes and Flying Boats of the Solent* Southampton University with Southampton City Museums, 1981

INDEX

Felixstowe, 33, 82, 89–91